2nd International Conference on Experimental and Computational Mechanics in Engineering (2nd ICECME 2020)

AA007697

Strengthening Applied Technology Innovation and Sustainable Development in Industry 4.0

IOP Conference Series: Materials Science and Engineering
Volume 1082

Banda Aceh, Indonesia
13 - 14 October 2020

Editors:

Mohammad Iqbal
Syarizal Fonna

ISBN: 978-1-7138-3347-5
ISSN: 1757-8981

Printed from e-media with permission by:

Curran Associates, Inc.
57 Morehouse Lane
Red Hook, NY 12571

Some format issues inherent in the e-media version may also appear in this print version.

This work is licensed under a Creative Commons Attribution 3.0 International Licence.
Licence details: http://creativecommons.org/licenses/by/3.0/.

No changes have been made to the content of these proceedings. There may be changes to pagination and minor adjustments for aesthetics.

Printed with permission by Curran Associates, Inc. (2021)

For permission requests, please contact the Institute of Physics at the address below.

Institute of Physics
Dirac House, Temple Back
Bristol BS1 6BE UK

Phone: 44 1 17 929 7481
Fax: 44 1 17 920 0979

techtracking@iop.org

Additional copies of this publication are available from:

Curran Associates, Inc.
57 Morehouse Lane
Red Hook, NY 12571 USA
Phone: 845-758-0400
Fax: 845-758-2633
Email: curran@proceedings.com
Web: www.proceedings.com

TABLE OF CONTENTS

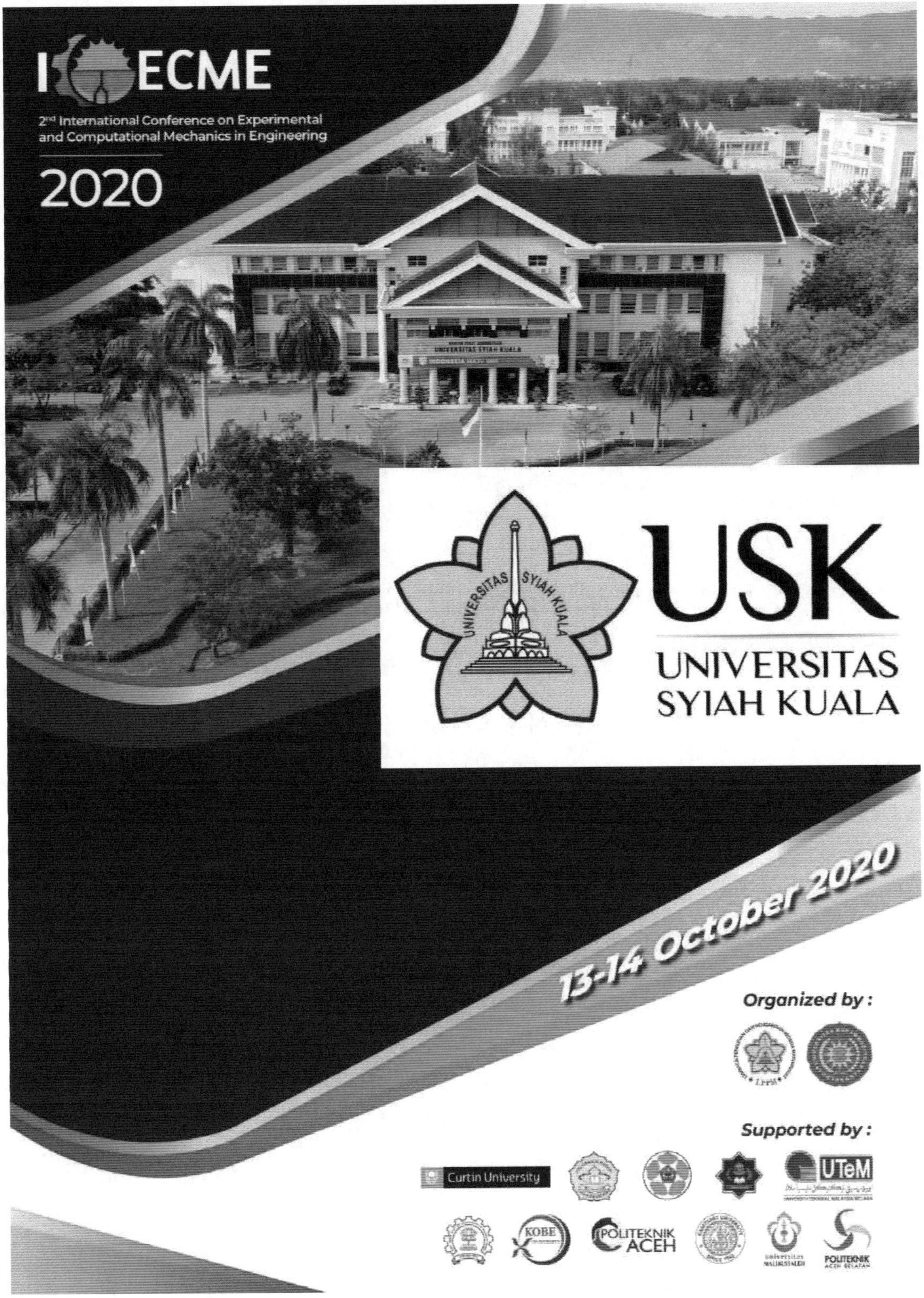

Content from this work may be used under the terms of the Creative Commons Attribution 3.0 licence. Any further distribution of this work must maintain attribution to the author(s) and the title of the work, journal citation and DOI.
Published under licence by IOP Publishing Ltd

IOP Conf. Series: Materials Science and Engineering 1082 (2021) 011001 doi:10.1088/1757-899X/1082/1/011001

2ND INTERNATIONAL CONFERENCE ON EXPERIMENTAL AND COMPUTATIONAL MECHANICS IN ENGINEERING - 2020 (ICECME - 2020)

DEPARTMENT OF MECHANICAL AND INDUSTRIAL ENGINEERING
FACULTY OF ENGINEERING, UNIVERSITAS SYIAH KUALA
OCTOBER 13-14, 2020

IOP Conf. Series: Materials Science and Engineering 1082 (2021) 011001 doi:10.1088/1757-899X/1082/1/011001

The 2th International Conference on Experimental and Computational Mechanics in Engineering (ICECME) 2020

Department Of Mechanical And Industrial Engineering

Faculty Of Engineering

Universitas Syiah Kuala

In conjunction with:
International Conference on Information Technology, Advanced Mechanical and Electrical Engineering (ICITAMEE) of Universitas Muhammadiyah Yogyakarta

IOP Conf. Series: Materials Science and Engineering 1082 (2021) 011001 doi:10.1088/1757-899X/1082/1/011001

Conference Information

Conference Name	International Conference on Experimental and Computational Mechanics in Engineering (ICECME) 2020
Conference Logo	
Host and Organizer	Dept. of Mechanical and Industrial Engineering, Faculty of Engineering, Universitas Syiah Kuala
Venue	ICT Centre, Universitas Syiah Kuala, Banda Aceh, Indonesia, and Virtual via Zoom Meetings
Official Language	English
Secretariat	Jurusan Teknik Mesin dan Industri, Fakultas Teknik, Universitas Syiah Kuala, Banda Aceh, Indonesia Phone +62 85361797661 Fax +62 651 7552222 Website http://conferencemechanic.unsyiah.ac.id/ Email icecme@unsyiah.ac.id

IOP Conf. Series: Materials Science and Engineering 1082 (2021) 011001 doi:10.1088/1757-899X/1082/1/011001

2nd ICECME 2020
International Conference on Experimental and Computational Mechanics in Engineering

Honorary Advisory Board

Rector of Universitas Syiah Kuala	Prof. Dr. Samsul Rizal
Vice Rector I	Prof. Dr. Marwan
Vice Rector II	Dr. Agussabti
Vice Rector III	Dr. Alfiansyah Yulianur
Vice Rector IV	Dr. Hizir
Head of Institute for Research and Community Services	Dr. Taufik Fuadi Abidin

Organizing Committee

Chairman	: Dr. Ir. Akhyar, M.P., M.Eng (Universitas Syiah Kuala, Indonesia)
Vice Chairman	: Dr.-Ing. Teuku Edisah Putra, S.T, M.Sc., Ph.D (Universitas Syiah Kuala, Indonesia)
Secretary	: Ratna Sary, S.T., M.T (Universitas Syiah Kuala, Indonesia)
Finance	: Dr. Ir. Razali, M.Si, M.T (Universitas Syiah Kuala, Indonesia)

Committee Members

1. Friesca Erwan, S.T., M.T., MProjMgt (Universitas Syiah Kuala, Indonesia)
2. Raihan Dara Lufika, S.T., M.Sc (Universitas Syiah Kuala, Indonesia)
3. Andriansyah, S.T., M.T. (Universitas Syiah Kuala, Indonesia)
4. Akram, S.T, M.T. (Universitas Syiah Kuala, Indonesia)
5. Amir Zaki Mubarak, S.T, M.T. (Universitas Syiah Kuala, Indonesia)
6. Ir. Asbar, M.T. (Universitas Syiah Kuala, Indonesia)
7. Ir. Jalaluddin, M.T. (Universitas Syiah Kuala, Indonesia)
8. Ir. Darwin Harun, M.T. (Universitas Syiah Kuala, Indonesia)
9. Ir. Masri Ali, M.Eng (Universitas Syiah Kuala, Indonesia)
10. Ir. Masri, M.T. (Universitas Syiah Kuala, Indonesia)
11. Muhammad Tadjuddin, S.T, M.Eng.Sc (Universitas Syiah Kuala, Indonesia)
12. Sabri, S.T, M.T. (Universitas Syiah Kuala, Indonesia)
13. Ir. Suhaeri, M.Eng (Universitas Syiah Kuala, Indonesia)
14. Syahriza, S.T, M.Eng (Universitas Syiah Kuala, Indonesia)
15. Teuku Firsa, S.T, M.Eng.Sc (Universitas Syiah Kuala, Indonesia)

16. Ir. Udink Aulia, M.Eng (Universitas Syiah Kuala, Indonesia)
17. Zulfadhli, S.T, M.T. (Universitas Syiah Kuala, Indonesia)
18. Zulfan, S.T, M.T. (Universitas Syiah Kuala, Indonesia)
19. Teuku Indra Rihayatsyah, S.T, M.Eng.Sc (Universitas Syiah Kuala, Indonesia)

Scientific Committees

1. Prof. Dr. Ir. Samsul Rizal M.Eng (Universitas Syiah Kuala, Indonesia)
2. Prof. Dr. Ir. Ahmad Syuhada, M.Sc (Universitas Syiah Kuala, Indonesia)
3. Prof. Dr. Ir. Husaini, M.T (Universitas Syiah Kuala, Indonesia)
4. Prof. Dr. Ir. Khairil, M.T (Universitas Syiah Kuala, Indonesia)
5. Prof. Dr.Ing Yulfian Aminanda (Universiti Teknologi Brunei, Brunei Darussalam)
6. Prof. Dr. T.M. Indra Mahlia, (Sydney Technology University, Australia)
7. Dr. Nitin Raut (Sohar University, Oman)
8. Ass. Prof. Dr. Fauzan Adziman (Science University of Oxford, United Kingdom)
9. Associate Professor Ir. Ts. Dr. Mohd Rizal Alkahari (Universiti Teknikal Malaysia Melaka (UTeM), Malaysia)
10. Professor Louis Cacetta (Curtin University, Australia)
11. Professor Chetan Singh Solanki (IIT Bombay, India)
12. Prof. Dr. Triyono, ST., MT. (Universitas Sebelas Maret-UNS)
13. Dr.Eng. Priyo Tri Iswanto, S.T., M.Eng. (Universitas Gadjah Mada-Indonesia-UGM)
14. Dr. Agus Suprihanto, ST., MT Dr. Agus Suprihanto, ST., MT (Universitas Diponegoro-UNDIP)
15. Ir. Viktor Malau, DEA., Ph.D (Universitas Gajah Mada-UGM-Indonesia)
16. Ubaidillah, Ph.D (Universitas Sebelas Maret-UNS-Indonesia)

IOP Conf. Series: Materials Science and Engineering 1082 (2021) 011001 doi:10.1088/1757-899X/1082/1/011001

Editorial Board

Chief of Editor:

Dr. Ir. Mohd. Iqbal, M.T. (Universitas Syiah Kuala, Indonesia)

Dr. Syarizal Fonna, S.T, M.Sc (Universitas Syiah Kuala, Indonesia)

Editors:

1. Dr. Irwansyah, ST, M.Eng (Universitas Syiah Kuala, Indonesia)
2. Dr. Sarwo Edhy S, S.T, M.Eng (Universitas Syiah Kuala, Indonesia)
3. Dr. Ir. Dipl. Ing. Nurdin Ali (Universitas Syiah Kuala, Indonesia)
4. Dr. Iskandar, S.T, M.Eng.Sc (Universitas Syiah Kuala, Indonesia)
5. Dr. Syifaul Huzni, S.T, M.Sc (Universitas Syiah Kuala, Indonesia)
6. Dr. Arhami, S.T., M.T (Universitas Syiah Kuala, Indonesia)
7. Dr. Husni, M.Eng.Sc (Universitas Syiah Kuala, Indonesia)
8. Dr. Ir. Muhammad Dirhamsyah, M.T (Universitas Syiah Kuala, Indonesia)
9. Dr. M. Nizar Machmud, S.T, M.Eng (Universitas Syiah Kuala, Indonesia)
10. Dr. Muhammad Rizal, S.T, M.Sc (Universitas Syiah Kuala, Indonesia)
11. Dr. Rudi Kurniawan, S. T, M.Sc. (Universitas Syiah Kuala, Indonesia)
12. Dr. Ir. Sulaiman Thalib, M.T (Universitas Syiah Kuala, Indonesia)
13. Dr. Zahrul Fuadi, S.T, M.Sc (Universitas Syiah Kuala, Indonesia)
14. Dr. Ir. Hamdani, M.T (Universitas Syiah Kuala, Indonesia)
15. Dr. Muhammad Ilham Maulana, S.T, M.T (Universitas Syiah Kuala, Indonesia)
16. Dr. Israr, S.T., M.Sc (Universitas Syiah Kuala, Indonesia)
17. Dr. Ikramullah, S.T. (Universitas Syiah Kuala, Indonesia)

Administrators

1. Milda Laura, SE (Universitas Syiah Kuala, Indonesia)
2. Fitri Yanti, SE (Universitas Syiah Kuala, Indonesia)

IOP Conf. Series: Materials Science and Engineering 1082 (2021) 011001 doi:10.1088/1757-899X/1082/1/011001

2nd ICECME 2020
International Conference on Experimental and Computational Mechanics in Engineering

About Unsyiah

Universitas Syiah Kuala (Unsyiah) is the oldest public university in Aceh. Standing on September 2, 1961 by Decree of the Minister of Higher Education and Science Number 11 of 1961, dated July 21, 1961. The establishment of Unsyiah confirmed by Decree of the President of Republic Indonesia, number 161 of 1962, dated April 24, 1962 in Kopelma Darussalam, Banda Aceh. Unsyiah domiciled in the capital of Aceh province with the main campus located in the city of Students (Kopelma) Darussalam, Banda Aceh. Currently, Unsyiah has more than 30,000 students studying in 12 faculties and the Graduate Program.

About ICECME

International Conference on Experimental and Computational Mechanic in Engineering is intended as an International Forum for those who wish to present their latest research results, innovative ideas, and experiences in the fields of Mechanical Engineering. It also offers the excellent opportunity to meet experts, exchanges information, and strengthen the collaboration among researchers, engineers, and scholars from academia, government, and industry, while enjoying the cultural heritage and natural beauty of Banda Aceh, coupled with the friendliness of its people.

Among the topics discussed in the ICECME 2020 were (but not limited to) Material Engineering, Mechanical Engineering, and Manufacturing Processing and System. ICECME has been conducted as annual international conference and will be continued in the years on wards. The organizer has been working seriously to manage the quality of the conference, as well as to increase the number of participant and the availability of the papers in a higher level of international publication.

IOP Conf. Series: Materials Science and Engineering 1082 (2021) 011001 doi:10.1088/1757-899X/1082/1/011001

WELCOME SPEECH FROM RECTOR OF UNIVERSITAS SYIAH KUALA

Assalamualaikum Wa Rahmatullahi Wa Barakatuh,In the name of Allah, the Most Beneficent and the Most Merciful. May peace, mercy, and blessings of Allah be upon you.

Dear colleagues, professors, lecturers, researchers, ladies and gentlemen. On behalf of Universitas Syiah Kuala, I would like to express my sincere gratitude and welcome you to the 2nd International Conference on Experimental and Computational Mechanic in Engineering (ICECME) 2020. The ICECME 2020 is held in conjunction with International Conference on Information Technology, Advanced Mechanical and Electrical Engineering (ICITAMEE) of Universitas Muhammadiyah Yogyakarta. Moreover, I honorably welcome our keynote speakers Professor Louis Cacetta from Curtin University, Australia, Professor Yoshikazu Nakai from Kobe University, Japan, Professor Chetan Singh Solanki from IIT Bombay, India, Associate Professor Ir. Ts. Dr. Mohd Rizal Alkahari from Universiti Teknikal Malaysia Melaka (UTeM), Malaysia. We also have with us moderator Associate Professor Anupun Terdwongworakul, Ph.D from Kasetsart University, Thailand.

It is my hope that the 2nd ICECME 2020 would be able to achieve its objective in providing an effective forum for academician, researchers, and practitioners to strengthen academics and industrial collaborations to enhance research and innovation. We believe that academics and industry must not work alone but must collaborate. Because with collaboration, the goals become easier, more effective, and efficient to achieve. And, we can change to a better environment and life.

We also believe knowledge, research and technology are for all people, for minorities, people with disabilities, and to help anyone. No matter how much we can accomplish by ourselves, whether it be research or development, it is never sufficient in this world of knowledge. Therefore, the focal drive of this conference is to exchange ideas, and by participating in this exchange, it is hoped that all parties who may benefit from the conference can apply it in managing activities in their areas. It is pleasing to note that the agenda of this conference covers a wide range of interesting topics related to all theoretical and practical aspects. There are Computational mechanics, Metallurgy and material science, Energy, Manufacturing processing systems, Industrial and system engineering, Biomechanics, Artificial intelligent, Micro/nano engineering, Micro-electro-mechanical system, Machine learning, Mechatronics, and Engineering design.

IOP Conf. Series: Materials Science and Engineering 1082 (2021) 011001 doi:10.1088/1757-899X/1082/1/011001

Last but not least, my deepest gratitude goes to the Advisory Board, Organizing Committee, International Scientific Committee, institutions, companies, and volunteers who have directly and indirectly supported the success of this seminar. The committee has organized a vibrant scientific program and is working hard to present highly respected and internationally notorious speakers to lead it. Although we try our finest to be professional, on behalf of Universitas Syiah Kuala, please accept our sincere apologies should there be inconveniences that occur before, during, or after the event. I wish you a very productive conference with exciting and encouraging discussions and exchange of knowledge so that together we can anticipate a future by strengthening academics and industrial collaborations to enhance research and innovation.

May God bless us all with good health to make this event a successful and enjoyable one!

Best Regards,

Prof. Dr. Ir. Samsul Rizal, M.Eng
Rector of Universitas Syiah Kuala

2nd ICECME 2020
International Conference on Experimental and Computational Mechanics in Engineering

WELCOME SPEECH FROM THE GENERAL CHAIR OF THE 2nd ICECME 2020

Assalamualaikum Wa Rahmatullahi Wa Barakatuh,

On behalf of the organizing committees, I would like to welcome all of you to the 2nd International Conference on Experimental and Computational Mechanic in Engineering (ICECME) 2020, held in Banda Aceh, Indonesia on 13 and 14 October 2020 by the Department of Mechanical and Industrial Engineering (JTMI), Faculty of Engineering, Universitas Syiah Kuala, Banda Aceh, Indonesia. The 2nd ICECME 2020 is held in conjunction with International Conference on Information Technology, Advanced Mechanical and Electrical Engineering (ICITAMEE) of Universitas Muhammadiyah Yogyakarta.

Although it is the second of ICECME, JTMI Department has previously conducted a series of similar conferences named Computational Mechanics and Numerical Analysis (CMNA). The first Regional Seminar of CMNA was held in 2002 in Banda Aceh. The Seminars were then annually held as the 2nd to 5th one from 2003 to 2006 in Banda Aceh. In 2008 the 6th CMNA become International Conference, jointly organized with the Department of Mechanical and Material Engineering, Universiti Kebangsaan Malaysia. The collaboration continued to the 7th International Conference of CMNA held in Banda Aceh in 2009. More than 200 papers from various institutions and countries have been published in the proceeding of CMNA. Furthermore, the Conference organized by JTMI began to change its name to ICECME which was held in 2019, this followed the development of variations of the topic of discussion during the conference.

Under the theme of the conference "Strengthening Applied Technology Innovation and Sustainable Development in Industry 4.0", this conference features rich four keynote speeches delivered by Professor Louis Cacetta from Curtin University, Australia, Professor Yoshikazu Nakai from Kobe University, Japan, Professor Chetan Singh Solanki from IIT Bombay, India, Associate Professor Ir. Ts. Dr. Mohd Rizal Alkahari from Universiti Teknikal Malaysia Melaka (UTeM), Malaysia. We also have with us moderator Associate Professor Anupun Terdwongworakul, Ph.D from Kasetsart University, Thailand. The 2nd ICECME 2020 offers a special opportunity to bring together professors, researchers and scholars around the globe, and serves as a platform to deliver innovative research results and latest trends and development in the fields of Computational mechanics, Metallurgy and material science, Energy, Manufacturing processing systems, Industrial and system engineering, Biomechanics, Artificial intelligent, Micro/nano engineering, Micro-electro-mechanical system, Machine learning, Mechatronics, and Engineering design.

All accepted papers in this conference have the opportunities to be published in reputable publication IOP Materials Science and Engineering Proceedings.

The conference has received more than one hundred and five papers have been accepted to be presented at the conference. The papers come from 20 institutions and 6 countries (Indonesia, Malaysia, Australia, India, Thailand, and Taiwan). To all members of the organizing committees, the international scientific committee, the reviewers, and the collaboration partners, we would like to thank all of them for their tremendous efforts to organize this conference successfully. Hopefully, the authors and participant could share their knowledge and experiences in research and industrial application and will come up with a new network and collaboration.

We look forward to having a successful conference, and we hope that all the attendees enjoy and benefit from this conference.

Best Regards,
Dr. Ir. Akhyar, MP., M.Eng
General Chair of the 2nd ICECME 2020

IOP Conf. Series: Materials Science and Engineering 1082 (2021) 011002 doi:10.1088/1757-899X/1082/1/011002

Peer review declaration

All papers published in this volume of *IOP Conference Series: Materials Science and Engineering* have been peer reviewed through processes administered by the Editors. Reviews were conducted by expert referees to the professional and scientific standards expected of a proceedings journal published by IOP Publishing.

- **Type of peer review: Single-blind**
  In the 2nd ICECME 2020 Conference, Scientific reviewers have reviewed the author's manuscript through a single-blind review mechanism. This means that the author does not know the identity of the reviewer. Two reviewers are selected according to their field of expertise for one manuscript.

- **Conference submission management system:**
  The conference management system used in the 2nd ICECME 2020 Conference is the free version of EasyChair.

- **Number of submissions received:** 26 papers.

- **Number of submissions sent for review:** 26 papers.

- **Number of submissions accepted:** 15 papers.

- **Acceptance Rate (Number of Submissions Accepted / Number of Submissions Received X 100):** (15/26) x 100 = 57.69 %.

- **Average number of reviews per paper:** 2 reviewer per paper

- **Total number of reviewers involved:** 17 reviewers

- **Any additional info on review process:** -

- **Contact person for queries:**
Dr. Akhyar (Department of Mechanical Engineering, Universitas Syiah Kuala, Jln. Syech Abdurrauf No.7 Darussalam – Banda Aceh, 23111, Indonesia, akhyar@unsyiah.ac.id)
Dr. Mohd. Iqbal (Department of Mechanical Engineering, Universitas Syiah Kuala, Jln. Syech Abdurrauf No.7 Darussalam – Banda Aceh, 23111, Indonesia, mohd.iqbal@unsyiah.ac.id)

Content from this work may be used under the terms of the Creative Commons Attribution 3.0 licence. Any further distribution of this work must maintain attribution to the author(s) and the title of the work, journal citation and DOI.
Published under licence by IOP Publishing Ltd

IOP Conf. Series: Materials Science and Engineering 1082 (2021) 012001 doi:10.1088/1757-899X/1082/1/012001

Differences in Pour Temperature Affect Hardness Properties of CuZn Brass Alloy through Metal Casting

Iqbal[1,2], Nurdin Ali[1,3], Husni Husin[1,4], Akhyar[1,3]*, Khairil[3], Ahmad Farhan[5]

[1]Doctoral Program, School of Engineering, Universitas Syiah Kuala, Banda Aceh, Indonesia, 23111
[2]Department of Mechanical Engineering, Universitas Abulyatama, Banda Aceh, Indonesia
[3]Department of Mechanical Engineering, Universitas Syiah Kuala, Jln. Syech Abdurrauf No.7 Darussalam – Banda Aceh, 23111, Indonesia
[4]Reaction Engineering and Catalysis Laboratory, Department of Chemical Engineering, Universitas Syiah Kuala, Banda Aceh, Indonesia, 23111
[5]Department of Physic Education, FKIP Universitas Syiah Kuala, Darussalam, Banda Aceh 23111, Indonesia

*E-mail: akhyar@unsyiah.ac.id

Abstract. The purpose of this study to evaluate the effect of pouring temperature on the Vickers hardness of CuZn brass alloys through metal casting. The pouring temperature varied from 1,060; 1,110; 1,160; 1,210; and 1,260 °C; while metal molds were not given preheat treatment or at room temperature conditions. The results obtained that the maximum hardness is 616 HV shown at 1,210 °C of the pouring temperature, while the minimum hardness is 574 HV can be seen at 1,210 °C of the pouring temperature. The significance of the difference in casting temperature on the hardness of CuZn brass alloy through metal casting was analyzed using one-way ANOVA. Generally, it appears that hardness increases with increasing pouring temperature. The effect of pouring temperature results in differences in solidifying rates and freezing intervals of brass alloys. Furthermore, it affects the grain structure growth of the brass alloy which led to the difference in Vickers hardness.

Content from this work may be used under the terms of the Creative Commons Attribution 3.0 licence. Any further distribution of this work must maintain attribution to the author(s) and the title of the work, journal citation and DOI.
Published under licence by IOP Publishing Ltd

1. Introduction

The application of brass alloys is very wide in the engineering field, Products that are frequently found such as bolts, nuts, electrical connectors, valve bodies, and hydraulic fittings. Brass alloys have the properties of excellent mechanical strength, high workability, and excellent corrosion and wear resistance [1]. Differences in pouring and mold temperatures can affect the cooling rate of metal alloys [2]. The solidification rate can affect the microstructure of the alloys so that it affects the hardness, tensile strength, percentage elongation, and impact toughness of the metal alloy. Increasing the cooling rate can increase the hardness, tensile strength, percentage elongation, and impact toughness of metal alloys, this can reduce the percentage of porosity and secondary dendrite arm spacing on metal alloys [3]. Research conducted to vary the effect of pouring temperature on the mechanical properties of metal alloys was evaluated in metal casting [4,5].

The hardness of brass alloys with copper (Cu) content of 65.493 wt.pct and zinc (Zn) 34.506 wt.pct and others have been investigated through metal casting, and the Briner hardness value is HB 110.44 kg/mm^2. [6]. Microstructural evolution investigations have been carried out on brass alloys produced by low-pressure die casting. It has been shown that the fine β-phase distribution increases the hardness of the alloy [7]. The result shows an increase in the hardness of the alloy from 62 to 156 BHN due to the addition of Sn. Studies on low-pressure die casting brass alloys have been carried out to predict casting defects [8]. The effect of lead content on the microstructure, hardness of cast bronze alloys in permanent molds has been observed. The results showed that the increase in hardness of the bronze alloy was influenced by the initial formation of the spinodal or modulated structure [9]. The hardness of the product shell wall thickness from the rapid casting in brass alloys has been evaluated using three-dimensional printing technology. As a result, the hardness variation obtained from casting brass alloys with shell mold thickness ranges from 84-102 VHN [10]. Several brass alloy castings have been reported for several cases, including shell mold [11], machining applications [12], different gate systems [13].

The microstructure, hardness, and machinability of the alloys have been studied on α-brass alloy [14]. The hardness and impact toughness of recycled metal alloys have been evaluated by the metal casting process [15]. The effect of casting temperature on the mechanical properties of metal alloys has been investigated during metal casting [16,17]. ANOVA test has been carried out to optimize the process parameters when squeeze casting of brass alloys [18]. Based on this background, the purpose of this study was to evaluate the effect of casting temperature on Vickers hardness of the brass alloy was evaluated using a one-way ANOVA statistical test.

2. Method and Parameters

CuZn brass alloy casting products with five variations of pouring temperature were used to evaluate the hardness behavior of the alloy. In this experiment, a gravity casting process was carried out to produce cast samples [19, 20, 21]. The metal casting process begins by cutting a cylindrical brass rod with a diameter of 25.4 mm and a length of 609.6 mm into small pieces for easy melting in the furnace. The melting furnace used is a furnace with LPG (liquefied petroleum gas) fuel with additional oxygen for complete combustion. The cast sample for hardness testing is produced in a medium-carbon steel metal mold in a rectangular shape, as shown in Figure 1.

The chemical composition of the brass alloy (CuZn) used in this experiment is as shown in Table 1. Figure 2 shows a binary phase diagram of a brass alloy at equilibrium. Referring to the phase diagram and chemical composition of the brass used in this experiment, it can be identified that the melting point is around 980 °C [22]. Then five variations of pouring temperature were selected, namely 1,060; 1,110; 1,160; 1,210; and 1,260 °C or liquidus + 80, liquidus + 130, liquidus + 180, liquidus + 230, and liquidus + 280 °C which were carried out in this experiment. Hopefully, the solidifying interval can affect the hardness of the cast-sample. The longer solidification interval provides a greater opportunity for the formation of metal alloy microstructure. Where the microstructure morphology of the alloy can affect the Vickers hardness value of the brass alloy.

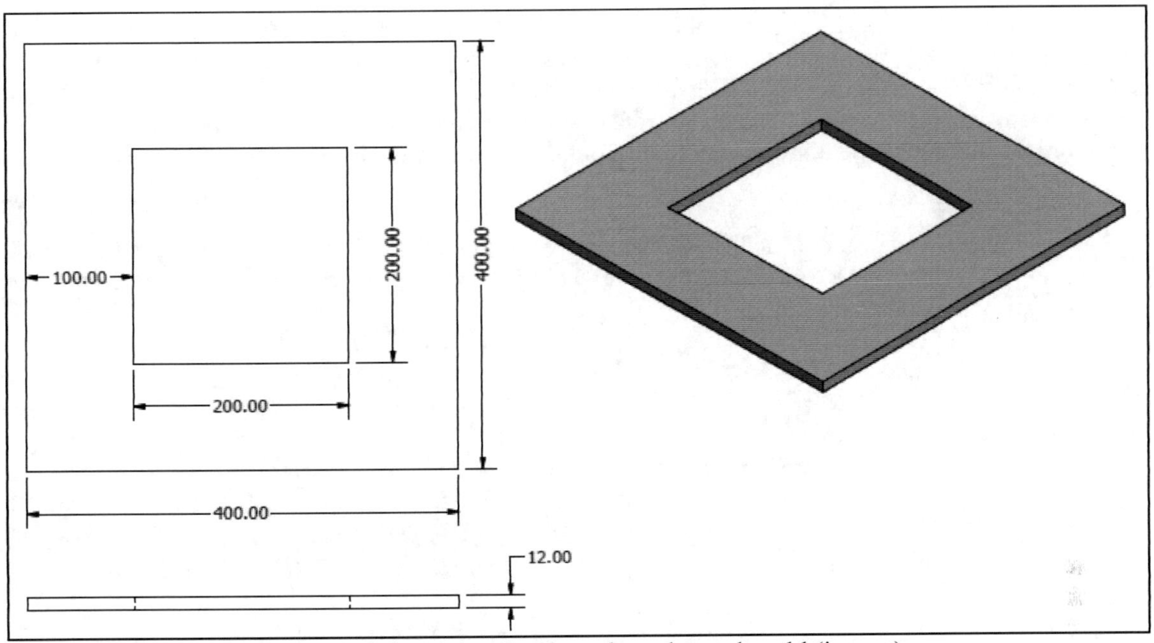

Figure 1. Schematic of a plate-shaped metal mold (in mm).

Table 1. Chemical composition of the brass alloy (wt.%).

Cu	Zn	Pb	Sn	Fe
Bal.	28.7	2.09	1.33	1.16

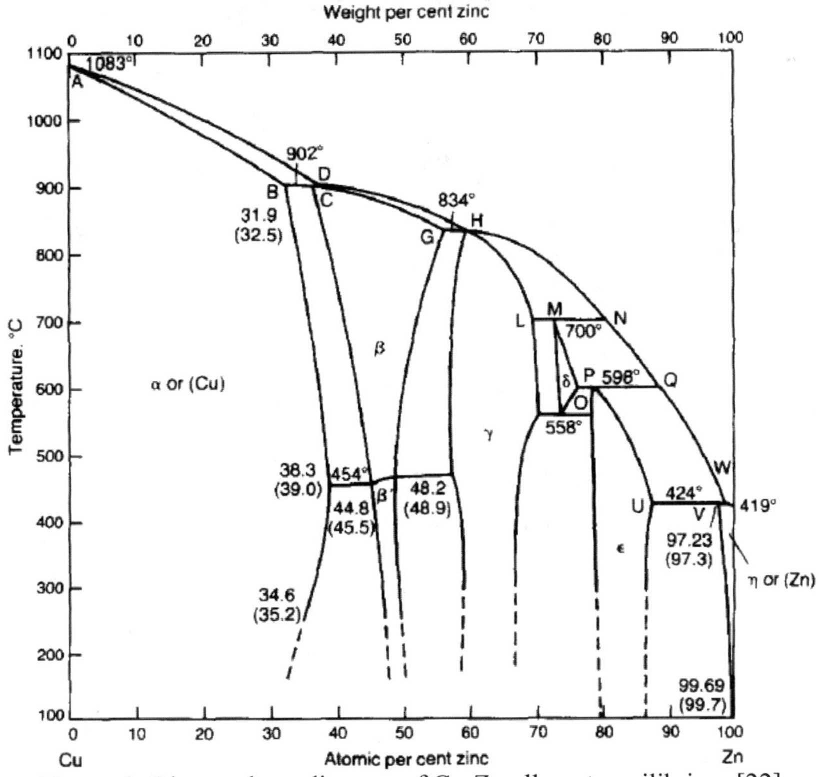

Figure 2. Binary phase diagram of Cu-Zn alloy at equilibrium [22].

IOP Conf. Series: Materials Science and Engineering 1082 (2021) 012001 doi:10.1088/1757-899X/1082/1/012001

The chemical composition test of the cast samples was carried out using the Spectrolab Jr CCD Spark Analyzer. The surface of the cast-sample was grinding and polished and observed through a scanning electron microscope (SEM). The Vickers hardness test on cast samples follows the ASTM E92 standard with ten test points each. Hardness analysis data obtained from experiments with five variations of pouring temperature were tested through one-way analysis of variance (ANOVA).

3. Result Analysis

The hardness value is one of the mechanical properties of an alloy which indicates the metal's resistance to indentation deformation. The brass alloy hardness value was evaluated against five variations of pouring temperature, as shown in Figure 3.

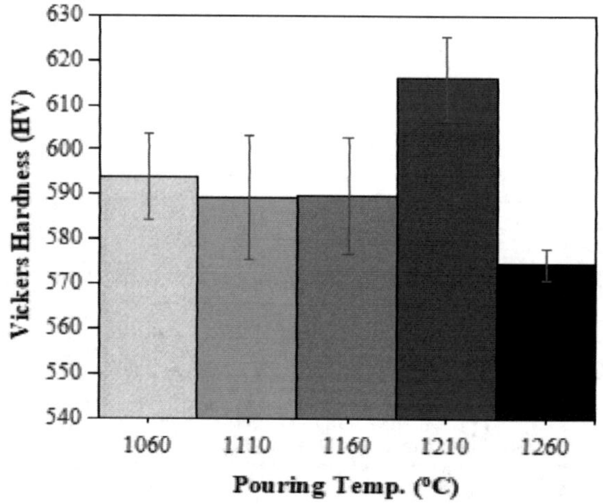

Figure 3. Vickers hardness of CuZn alloy with five variations of pouring temperature.

The results of the Vickers hardness test are shown in Figure 3, that differences in pouring temperature can affect the Vickers hardness of cast brass alloys. In general, based on the trend line, the hardness decreases with increasing pouring temperature. The minimum hardness value is 574 HV obtained from the pouring temperature at 1,260 °C, while the maximum hardness value is 616 HV obtained from the pouring temperature at 1,210 °C. The hardness value of the cast-sample in this experiment was higher when compared to the thick-shell mold of the brass alloy which had a value of 102 VHN [10].

The solidification interval can affect the growth of metal grains in metal alloys. A higher pouring temperature results in a larger solidification interval and causes a longer freezing time. High-temperature pouring can make the grain size enlarge and the phase distribution grows more evenly in the phase matrix. This is almost the same as the sintering conditions of Cu-Zn brass alloys, where the sintering time causes grain growth which causes the number of liquid phases, and the adjustment of the atomic diffusion conditions [23]. The alloy surface morphology was observed through SEM imaging with 500x magnification against five differences in casting temperature during the metal casting process as shown in Figure 4.

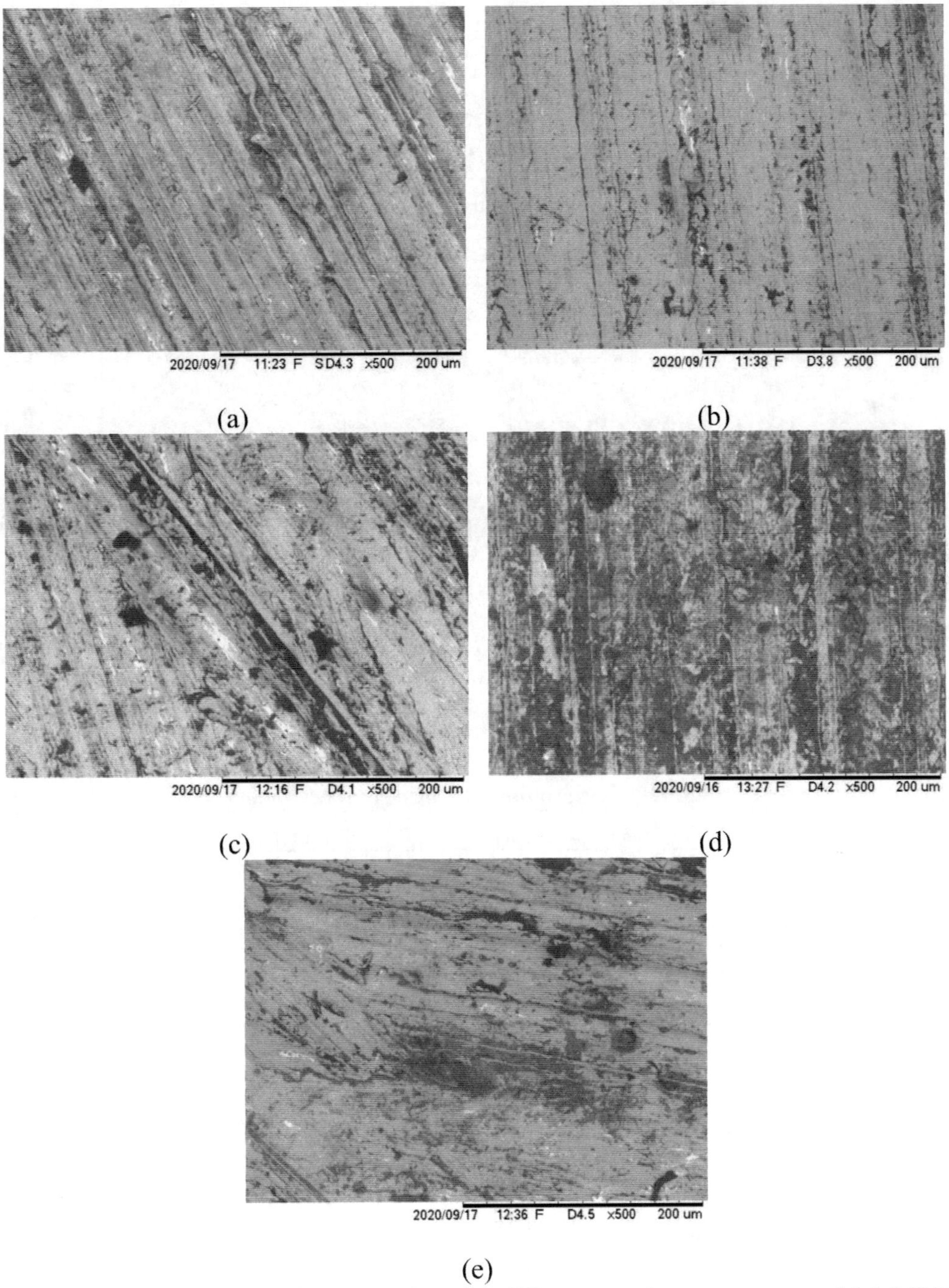

Figure 4. Surface morphology of cast samples with different pouring temperatures: (a) 1,060 °C; (b) 1,110 °C, (c) 1,160 °C, (d) 1,210 °C, and (e) 1,260 °C.

The hypothesis in this experiment was tested using one-way ANOVA. Homogeneity and normality tests were performed as statistical prerequisite tests. It was found that the subjects with different pouring temperatures came from normally distributed populations and homogeneous groups of Vickers hardness values (Table 2).

Table 2. Results of the hardness experiment.

No.	Pouring Temperature (°C)				
	1,060	1,110	1,160	1,210	1,260
1	595.27	595.27	588.58	657.03	664.92
2	591.91	598.66	608.99	634.18	657.03
3	660.96	578.76	572.35	626.83	572.35
4	605.52	582.01	689.47	697.96	626.83
5	602.07	702.26	578.76	702.26	572.35
6	575.54	657.03	582.01	657.03	752.32
7	619.60	608.99	689.47	637.90	894.29
8	591.91	572.35	776.87	605.52	578.76
9	595.27	818.71	598.66	612.50	1309.51
10	637.90	980.84	598.66	619.60	697.96

Table 3 shows the calculation results of sample number for each pouring temperature (*Count.*), The total Vickers hardness value of all samples per pouring temperature (*Sum*), the average Vickers hardness value of all samples per pouring temperature (*Mean*), the variance of the Vickers hardness value of all samples per pouring temperature (*Variance*), the sum of squares of the Vickers hardness value of all samples per pouring temperature (*SS*), standard error of each pouring temperature (*Std. Err.*), lower and upper limits of each pouring temperature. Table 4 shows the results of the one-way ANOVA calculation including the Mean Square (*MS*) value between Group, Within Group and Total, degrees of freedom (*df*).

Table 3. Description of hardness experiments.

Groups **(°C)**	***Count.***	***Sum***	***Mean***	***Variance***	***SS***	***Std Err***	***Lower***	***Upper***
1,060	10	6075.95	607.60	640.94	5768.45	38.6984	520.0536	695.1372
1,110	10	6694.87	669.49	17771.44	159942.97	38.6984	581.9450	757.0285
1,160	10	6283.81	628.38	4554.64	40991.79	38.6984	540.8388	715.9224
1,210	10	6450.80	645.08	1125.59	10130.33	38.6984	557.5378	732.6214
1,260	10	7326.31	732.63	50785.56	457070.02	38.6984	645.0889	820.1725

Table 4. ANOVA calculation results.

Sources	***SS***	***df***	***MS***	***F***	***P value***	***F crit***
Between Groups	92772.38	4	23193.0954	1.548722042	0.204424772	2.578739184
Within Groups	673903.56	45	14975.6346			
Total	766675.94	49	15646.4477			

Conclusions can be built from the calculated *F* value whether greater or less than the *F* table. If *F* Count < *F* Table, then H_0 is accepted, which means that the influence of pouring temperature on the Vickers hardness value is not significant at the critical alpha limit α (> 0.05). Conversely, if *F* Count > *F* Table then H_0 is rejected, which means that the effect of pouring temperature on the hardness value is significant at the critical alpha limit α (< 0.05). In the results of this experiment, the hardness value shows a significant probability at $\alpha < 0.3$ to variations in pouring temperature. From the ANOVA data, the calculated *F* value is greater than the *F* table (2.5787 > 1.5487) and it can be concluded that the increase of the pouring temperature can reduce the Vickers hardness value when casting brass alloys.

4. Conclusions

Vickers hardness evaluation on brass alloys with five variations of pouring temperature has been carried out in this experiment with the metal casting process. Based on the results and discussion, it can be concluded that variations in pouring temperature can affect the hardness value of brass alloys. The Vickers hardness value appears to decrease with increasing pouring temperature. The maximum Vickers hardness value is 616 HV obtained in cast samples with a pouring temperature of 1,210 °C, while the minimum value is 574 HV at a pouring temperature of 1,260 °C. The higher pouring temperature causes the cooling rate (time) to be wider, thus causing the grain growth to be larger than the rapid solidification time. From the ANOVA test, it can be concluded that there is a significant influence between the pouring temperature and the Vickers hardness value.

Acknowledgments
This publication was made possible by a grant from the Research Fund of the Ministry of Research, Technology, and Education, Republic of Indonesia, and LPPM Universitas Syiah Kuala; the financial support is greatly appreciated. We would like to thank Mr. Samsul Bahri for their support during this experiment.

Reference

[1] Toulfatzis A I, Pantazopoulos G A, and Paipetis A S, 2014, *Fracture Behavior and Characterization of Lead-Free Brass Alloys for Machining Applications*, Journal of Materials Engineering and Performance, **23**, 3193–3206, DOI: 10.1007/s11665-014-1096-3

[2] Akhyar H, Husaini, 2016, *Study on Cooling Curve Behavior During Solidification and Investigation of Impact Strength and Hardness of Recycled Al–Zn Aluminum Alloy*, International Journal of Metalcasting, **10**, 452–456, https://doi.org/10.1007/s40962-016-0024-8

[3] Savaskan T, and Turhal M S, 2003, *Relationships between cooling rate, copper content and mechanical properties of monotectoid based Zn–Al–Cu alloys*, Materials Characterization, **51**, 259– 270, https://doi.org/10.1016/j.matchar.2003.12.003

[4] Akhyar H, Iswanto P T, and Malau V, 2018, *Non Treatment, T4 and T6 on Tensile Strength of Al-5.9Cu-1.9Mg Alloy Investigated by Variation of Casting Temperature*, Materials Science Forum, **929**, 56-62, doi:10.4028/www.scientific.net/MSF.929.56.

[5] Iswanto P T, Akhyar, Pambekti A, 2020, *Heat treatment T4 and T6 effects on mechanical properties in Al-Cu alloy after remelt with different pouring temperatures*, Metalurgija, **59(2)**,171-174

[6] Turchi P E, Sluiter M, Pinski F J, Johnson D D, Nicholson D M, Stocks G M, and Staunton J B, 1991, *First-principles study of phase stability in Cu-Zn substitutional alloys*, Physical Review Letters, **67(13)**,1779-1782, doi: 10.1103/PhysRevLett.67.1779

[7] Yang C, Ye W, Yang Y, Wang F, Ding Y, Zhu Q, and Zhang W, 2018, *Microstructural Evolution and Mechanical Behavior of Lead-Free Silicon Brass Manufactured by Low-Pressure Die Casting*, Journal of Materials Engineering and Performance, **27**, 5478–5488, https://doi.org/10.1007/s11665-018-3615-0

[8] Muikku A, Hartikainen J, Vapalahti S, Tiainen T, 2006, *Experimental Work on Possibilities to Predict Casting Defects in LPDC Brass Castings*, Materials Science Forum, **508**, 561-566, doi:10.4028/www.scientific.net/MSF.508.561

[9] Shankar K V, and Sellamuthu R, *Determination on The Effect of Tin Content on Microstructure, Hardness, Optimum Aging Temperature and Aging Time for Spinodal Bronze Alloys Cast in Metal Mold*, International Journal of Metalcasting, **11**, 189–194, DOI 10.1007/s40962-016-0034-6

[10] Singh R, and Singh J P, 2009, *Comparison of rapid casting solutions for lead and brass alloys using three-dimensional printing*, Proceedings of the Institution of Mechanical Engineers, Part C: Journal of Mechanical Engineering Science, **223**, 2117-2123, DOI: 10.1243/09544062JMES1387

[11] Chhabra M, and Singh R, 2015, Mathematical Modeling of Surface Roughness of Castings Produced Using ZCast Direct Metal Casting, Journal of The Institution of Engineers (India): Series C, **96**, 145–155, DOI 10.1007/s40032-014-0139-8
[12] Toulfatzis A I, Pantazopoulos G A, and Paipetis A S, 2014, *Fracture Behavior and Characterization of Lead-Free Brass Alloys for Machining Applications*, Journal of Materials Engineering and Performance, **23**, 3193–3206,, DOI: 10.1007/s11665-014-1096-3
[13] Akhyar, Husaini, Ali M, Ali N, and Farhan A, 2020, *Effect of Different Gating Systems and Sand Mold Binder on the Cast-Quality of Bicycle Frame Produced through Sand Casting Method*, Defect and Diffusion Forum, **402**, 100-107, https://doi.org/10.4028/www.scientific.net/DDF.402.100
[14] Rajabi Z, and Doostmohammadi H, 2018, *Effect of addition of tin on the microstructure and machinability of α-brass*, Materials Science and Technology, **34**, 1218-1227, https://doi.org/10.1080/02670836.2018.1435484
[15] Farhan A, Akhyar, and Ali M, 2019, *Experiment on Hardness and Impact Strength of Recycled Aluminum Alloys by Metal Casting Process*, Materials Science Forum, **961**, pp 65-72, https://doi.org/10.4028/www.scientific.net/MSF.961.65
[16] V. Malau, Akhyar, P. Tri Iswanto, 2018, Modification of Constrained Rod Casting Mold for New Hot Tearing Measurement, Archives of Metallurgy and Materials, **63(3)**, 1201-1208, DOI: 10.24425/123792
[17] Akhyar H, Malau V, Suyitno, Iswanto P T, 2017, *Hot tearing susceptibility of aluminum alloys using CRCM-Horizontal mold*, Results in Physics, **7**, 1030–1039, http://dx.doi.org/10.1016/j.rinp.2017.02.041
[18] Singh D, Chhabra M, and Tirth V, 2015, *Optimization of Process Parameter to Enhance the Impact Strength of Squeeze Cast Brass Alloy*, MIT International Journal of Mechanical Engineering, **5(1)**, 43-48,
[19] Farhan A, Akhyar, and Ali M, 2019, Experiment on Hardness and Impact Strength of Recycled Aluminum Alloys by Metal Casting Process, Materials Science Forum, 961, 65-72, doi:10.4028/www.scientific.net/MSF.961.65
[20] Iswanto P T, Akhyar, Maliwemu E U K, 2019, Fatigue Crack Growth Rate of Motorcycle Wheel Fabricated by Centrifugal Casting, Metalurgija, 58(1-2), 51-54.
[21] Iswanto P T, Akhyar H, Malau V, Suyitno, and Wirawan R, 2016, Effect of T6 heat treatment and artificial aging on mechanics and fatigue properties of A356.0 aluminum alloy produced by 350 rpm centrifugal casting, 2016 6th International Annual Engineering Seminar (InAES), Yogyakarta, 2016, 223-227, doi: 10.1109/INAES.2016.7821938.
[22] Ahlers M, 1986, *Martensite and equilibrium phases in Cu-Zn and Cu-Zn-Al alloys, Progress in Materials Science*, **30(3)**, 135-186, https://doi.org/10.1016/0079-6425(86)90007-1
[23] Hasani A, Azadbeh M, and Moradi M, 2020, *Sintering Process of Cu–28Zn Brass Alloy: Statistical Investigation*, Transactions of the Indian Institute of Metals volume, **73**, 1383–1400, https://doi.org/10.1007/s12666-020-01982-0

IOP Conf. Series: Materials Science and Engineering 1082 (2021) 012002 doi:10.1088/1757-899X/1082/1/012002

Optimum Load in Vickers Microhardness Testing of Treated Magnesium Alloy by Thermal Oxidation

Lutiyatmi[1]*, Sutiyoko[1], Ridwan Afandi[1]

[1]Foundry Department of Politeknik Manufaktur Ceper, Klaten, Indonesia

*E-mail: yatmiluti@gmail.com

Abstract. Magnesium and its alloy have good properties as biomedical implant such as biodegradable material and excellent biocompatibility. Their hardness is an important mechanical property in using of these materials. Vickers microhardness is one of hardness testing methods in metal materials especially for thin layer on the metal. Mass of microhardness testing load also has important effect on result validity. The object of this research is to investigate optimum load in Vickers microhardness testing of magnesium alloy treated by thermal oxidation. Magnesium alloy AZ61 was treated by annealing at 150; 225; 300; 375; 450 °C of temperature. Vickers microhardness testing was carried out on this treated material using 5, 10, 25, 50, and 100 g of load. Five data for both each load and temperature, therefore there are 25 data for each load. Optimum load is determined the load which has lowest standard deviation in Vickers microhardness. The optimum load in this research was happened at 10 g in load of Vickers microhardness testing. The Vickers microhardness in this load was 36.0 g.mm^{-2} and the standard deviation was 2.7 g.mm^{-2}.

Content from this work may be used under the terms of the Creative Commons Attribution 3.0 licence. Any further distribution of this work must maintain attribution to the author(s) and the title of the work, journal citation and DOI.
Published under licence by IOP Publishing Ltd

1. Introduction

Development of biomaterial for fracture fixation or joint replacement has an important role in orthopedic surgery. Orthopedic biomaterial is implanted in body to heal of the fracture bone [1]. Magnesium and it alloy is one of the biomaterial which have some advantages as an implant material. Magnesium already existed more than one century [2] and has been improved until now. This material has density (1.738 g/cm^3) [1] more closely to cortical bone density (1.75 – 2.1 g/cm^3) [3] than other metal implants. Another advantage, its biocompatibility is good and increase the bone formation in the body [4]. Beside of their advantages, the corrosion resistance is one of the disadvantages of these materials.

Treatment of these materials have been developed to reduce their drawbacks. Surface treatments such as coating, shot peening, surface mechanical attrition treatment already were developed to minimize their weakness. Coating treatment can be done by thermal oxidation, ion implantation, etc. Ion reaction between substrate and its environment [5] or deposit the coating material on the substrate surface [6] are common method in coating treatment. Coating of magnesium by embedded in NaOH solution results in passive layer on the substrate surface [7]. Corrosion resistance of magnesium enhance by coating of calcium phosphate [8]. Coating by fluoride on magnesium formed MgF_2 and also increased the corrosion resistance [9]. Biodegradable of coated magnesium by phosphate coating is lower than that in original material [10].

Thermal oxidation is a simple method in treatment of material [11]. Thermal oxidation of magnesium alloy AZ91 results in oxide layer composing pile of oxide grains covered entirely of magnesium surface [12]. Thermal oxidation treatment changes mechanical properties of materials such as tensile strength, hardness, toughness. The hardness change occurred on the top surface to the inner materials. One of the methods to measure hardness of treated material by thermal oxidation is Vickers microhardness. Both excessive and lack of the load in Vickers microhardness testing will effect in data validity. The aim of this investigation is to achieve optimum load in Vickers microhardness testing of magnesium alloy AZ61 treated by oxidation thermal.

2. Method and Parameters

Magnesium alloy AZ61 specimen was prepared in 20×20×6 mm^3 of dimension. There are five specimens for five temperatures (150; 225; 300; 375; 450 °C) in thermal oxidation. Specimens were prepared by polishing with abrasive paper until 1500 mesh. Thermal oxidation used Nabhertham heat treatment furnace which has 1200 °C of maximum temperature. Each specimen was annealed in this furnace for 6 hours at each temperature. Vickers microhardness test used Vickers microhardness machine (FM-800, Future-Tech Corp.). Load parameters were arranged at five loads (5, 10, 25, 50, and 100 g) for each specimen and five repetitions for each parameter. Vickers Hardness Number (VHN) for each load was calculated by average of 25 data and its standard deviation also calculated from 25 data. The optimum load was chosen at the least standard deviation.

3. Result Analysis

Vickers hardness and its standard deviation for each load at each thermal oxidation temperature are shown in Table 1. The first row demonstrate the loads from 5 to 100 g and the first column shows the thermal oxidation temperature form 150 to 450 °C. At 150 °C, the Hardness Vickers (HV) is 34.5 for 5 g of load. The HV increases with the load and achieves maximum HV at 42.4 for 25 of load, after that the HV decreases with load increase. At 225 °C, the HV decreases from 37.1 to 35.8 for 5 g to 10 g of load and increase until 47.6 at 50 g of load. For 300 °C of thermal oxidation, the HV increases as load increase and achieve the maximum HV at 38.0 for 100 g of load. The HV increases from 22.5 to 36.2 for 5 – 50 g of load at the 375 °C of thermal oxidation and decreases to 27.2 at 100 g of load. For the higher temperature of thermal oxidation (450 °C), the HV is inconsistent result. It increases from 5 to 10 g of load, decreases for 25 g of load, increases at 50 g of load and decreases again at 100 g of load.

Table 1. Vickers Hardness of magnesium alloy treated by thermal oxidation

Temperature (°C)	Load : 5 g		Load : 10 g		Load : 25 g		Load : 50 g		Load : 100 g	
	HV ($g.mm^{-2}$)	STD EV	HV ($g.mm^{-2}$)	STD EV	HV ($g.mm^{-2}$)	STD EV	HV ($g.mm^{-2}$)	STD EV	HV ($g.mm^{-2}$)	STD EV
150	34.5	2.5	36.3	2.4	42.4	11.9	39.7	7.2	34.2	1.6
225	37.1	4.0	35.8	1.8	43.4	5.0	47.6	4.6	39.2	1.8
300	27.3	1.9	34.2	2.4	35.3	1.5	35.9	0.9	38.0	3.1
375	22.5	1.5	29.6	1.8	31.8	1.9	36.2	3.7	27.2	6.5
450	39.8	8.2	44.0	5.3	37.8	6.1	50.1	6.5	46.4	2.2
Average	**32.2**	**3.6**	**36.0**	**2.7**	**38.1**	**5.3**	**41.9**	**4.6**	**37.0**	**3.1**

Figure 1 represents the HV of magnesium alloy treated by thermal oxidation for load variation. The HV for 5; 10; 25; 50; 100 g of load are 32.2 ± 3.6; 36.0 ± 2.7; 38.1 ± 5.3; 41.9 ± 4.6; 37.0 ± 3.1 respectively. The HV increases from 32.2 to 41.9 for 5 to 50 g of load and decreases at 100 g of load. Deviation standard of 25 g load is 5.3 and it is a highest standard deviation in this research. The lowest standard deviation is 2.7 at 10 g of load. The average HV for all loads is 37.0 and the standard deviation is 3.9.

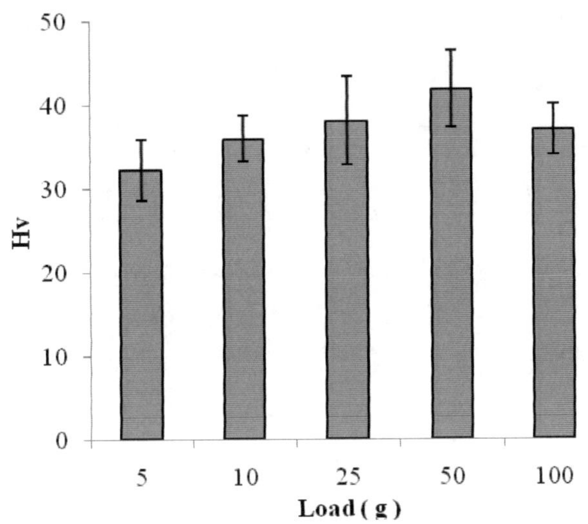

Figure 1. Vickers microhardness of magnesium alloy treated by thermal oxidation

Vickers microhardness of magnesium alloy has variety value due to the load of Vickers microhardness testing. The optimum load is chosen base on the deviation standard of results. The least of standard deviation indicates more consistent of the value. The load of 10 g has the least standard deviation, therefore the optimum load in this Vickers microhardness testing is 10 g.

4. Conclusions

One of the Vickers microhardness test parameter is the load in indentation. Optimum load should be chosen to get better result in Vickers microhardness. The optimum load in this research is 10 g because of the least standard deviation in the Vickers Hardness. Higher standard deviation makes the results have less accuracy.

Acknowledgments
This work was supported by Indonesian Ministry of Research in fundamental research grant.

Reference

[1] Radha R, Sreekanth D, 2017, *Insight of magnesium alloys and composites for orthopedic implant applications – a review, Journal of Magnesium Alloy.*, **5**, 286–312.

[2] Witte F, Abeln I, Switzer E, Kaese V, Meyer-Lindenberg A, Windhagen H, 2008, *Evaluation of the skin sensitizing potential of biodegradable magnesium alloys*, Journal of Biomedic Materials Research part A, **86(4)**, 1041–1047.

[3] Richards AM, Coleman NW, Knight TA, Belkoff SM, Mears SC, 2010, *Bone Density and Cortical Thickness in Normal, Osteopenic, and Osteoporotic Sacra*, Journal of Osteoporosis, 1–5.

[4] Lopez HY, Cortes-Hernandez DA, Escobedo S, Mantovani D, 2006, *In Vitro Bioactivity Assessment of Metallic Magnesium*, Key Engineering Materials., **309–311**, 453–456.

[5] Hornberger H, Virtanen S, Boccaccini AR, 2012, *Biomedical coatings on magnesium alloys – A review*, Acta Biomaterialia, **27(2)**, 153–164.

[6] Song G, 2010, *Corrosion behavior of magnesium (Mg) alloys*, in *Surface engineering of light alloys*, H. Dong, Ed. Cambridge, UK: Woodhead Publishing Ltd, 3–39.

[7] Pourbaix M, 1974, *Atlas of electrochemical equilibrium in aqueous solutions*, Second. Oxford: Pergamon Press Ltd.

[8] Hiromoto S, Yamamoto A, 2009, *High corrosion resistance of magnesium coated with hydroxyapatite directly synthesized in an aqueous solution*, Electrochimica Acta, **54(27)**, 7085–7093.

[9] Supplit R, Koch T, Schubert U, 2007, *Evaluation of the anti-corrosive effect of acid pickling and sol–gel coating on magnesium AZ31 alloy, Corrosion Science*, **49(7)**, 3015–3023.

[10] Yang JX, Cui FZ, Lee IS, Zhang Z, 2011, *In vivo biocompatibility and degradation behavior of Mg alloy coated by calcium phosphate in a rabbit model*, Journal of Biomaterials Application, **1**, 153-64.

[11] Hanzi AC, Gunde P, Schinhammer M, Uggowitzer PJ, 2008, *On the biodegradation performance of an Mg–Y–RE alloy with various surface conditions in simulated body fluid, Acta Biomaterialia*, **5(1)**, 162–171.

[12] Majumdar JD, Bhattacharyya U, Biswas A, Manna I, 2008, *Studies on thermal oxidation of Mg-alloy (AZ91) for improving corrosion and wear resistance*, Surface Coating Technology, **202**, 3638–3642.

IOP Conf. Series: Materials Science and Engineering 1082 (2021) 012003 doi:10.1088/1757-899X/1082/1/012003

Innovative And Ergonomic Design of Baby Stroller for Children's 6 Months - 3 Years Old

H Soewardi [1], A Nariswari [1]
[1]Industrial Engineering Department, Universitas Islam Indonesia, Yogyakarta, Indonesia

E-mail: hartomo@uii.ac.id

Abstract. Baby stroller is a transportation device for carrying a toddler easily and safely. Nowadays, many types of stroller were developed and used by people. However, as much as 54.5% of users experienced dissatisfaction in utilizing the stroller which is difficult to climb stairs, damaged easily, unpractical, difficult to fold, stored goods fall easily, heavy, and takes up space. Thus, it is significance to improve this stroller design to be more ergonomic. Objective of this study is to develop an innovative design of stroller on basis ergonomic principles for alleviating some problem in use. Integration of axiomatic design method and house of quality concept was implemented in this study to determine the specification. Anthropometric data of users is also applied into the design. Non-parametric statistic method was applied to test the hypotheses. The results of this study is an innovative design of stroller in which it is equipped with chain wheels, handbrakes, and phosphor sticker on the wheels as well as it is more comfortable, safer, and more efficient. This new design is also valid to satisfy user requirement and differences with the existing design at 5% of significance level.

1. Introduction

Baby stroller is a transportation device used by people to carry a toddler during their activities. The idea of this device is necessity for parents to do many task in everyday life easily and comfortably [1] such that they do not need to hold their child who is going in the growth period. According to [2], this devices has also become the serious attention for some researchers in the world. They tried to find or develop the design of the stroller for supporting the people activity while keeping the baby securely.

Many kinds of the stroller is in the market from a simple design up to an advance design used to help people in carrying a baby. On the other, some innovation of stroller design has not been changed during 1980-2014 period [3]. Likewise, in China, the baby stroller manufacturers had not still focused on the functional innovation but they prioritized on the shape and appearance of the frame [4]. However, the existing stroller developed has not still be able to accommodate the user need. It is indicated, based on a preliminary study, that as much as 54.5% of user was not satisfied such as difficult to climb stairs, damaged easily, unpractical, difficult to fold, stored goods fall easily, heavy, and takes up space [5]. In general, users need to hold their child first then fold and lift the stroller thereby users feel hassles and fatigue. This is certainly very difficult for parents especially when he is done it all by himself and still carry baby equipment. Occasionally they keep a child's bag or equipment, in the stroller without thinking about the safeness, especially for items that contain liquid

Content from this work may be used under the terms of the Creative Commons Attribution 3.0 licence. Any further distribution of this work must maintain attribution to the author(s) and the title of the work, journal citation and DOI.
Published under licence by IOP Publishing Ltd

thereby it is possible for children to be spilled from those items [6]. Thus, it is significance to redesign the existing baby stroller to be better.

Some research has been done to design the stroller. They are baby car seats that has functioned as a baby stroller [7]. Another research developed a baby stroller that was functioned as a motion chair used to put a baby to sleep [8]. While another baby stroller had been designed with adding removable and reversible seat [9]. The safeness baby stroller was developed with automatic braking system and night time visibility [10].

The objective of the research study is to redesign the innovative baby stroller based on ergonomic concept to increase the comfort and safe in use for people by applying integration between axiomatic design method and house of quality.

2. Research Method

2.1. Survey
Paper based survey was conducted to identify users attributes or requirements of the design by distributing questionnaire to more than 90 respondents. They are people who have experience in using a baby stroller. Their ages ware in range from 16 - 55 years old.

2.2. Direct Measurement
Direct measurement using anthropometer was carried out to determine the anthropometric dimensions of stroller. It evolved 30 adults whose ages were 16 to 55 years old and 30 children whose ages were 6 month – 3 years old. Anthropometric data of the adults consist of Standing Elbow Height, Width of Index Finger, and Maximum Hand Grip Diameter. As for body dimension of encompasses Upright Sitting Height, Shoulder Width, Sitting Elbow Height, Popliteal Height, Buttock Popliteal, and Foot Length.

2.3. Axiomatic Design Method
The axiomatic design method was developed by Professor Nam Pyo Suh from Massachusetts Institute of Technology in 1970s as an effort to make the logic of the design process. This method is based on two axioms that is the independent axiom and the information axiom. The independent axiom states that the independence of FRs must be maintained for robustness, simplicity, and reliability of systems. While the information axiom states that the system must be designed to minimize the information content that is to maximize the probability of success in achieving the FRs [12]. Subsequently Suh explains several important terms in this method especially to the independent axiom, there are [11]:

CA : Customer Attribute. The domain that is used to accommodate the needs from the user's point of view.

FR : Functional Requirement. The domain that is used to accommodate the functions to be achieved from a design or product.

DP : Design Parameter. The domain that is the interpretation of the FR or how the function of the FR domain is realized.

PV : Process Variable. The domain that discusses how a design or product is produced or PV is the domain of the production process from a design before it becomes a product.

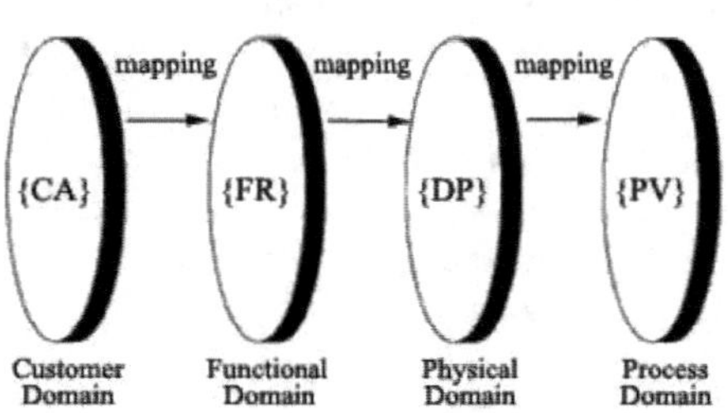

The Figure 1, shows the process design from the costumer's needs to the manufacturing process required for the product. The costumer's needs are listed is the Costumer Attribute Domain. These attributes are translated into functional requirement within the Functional Domain. Design parameters

Figure 1. Mapping Process Axiomatic Design Method [13].

are created to resolve each functional requirements. These design parameters create are physical design within the Physical Domain. Finally, process variables are determined in the Process Domain for the manufacturing of the product.

2.4. Integration of Axiomatic Design and House of Quality Method (AHOQ)
Integration of Axiomatic Design and House of Quality Method (AHOQ) is a proposed method in making modifications on HOQ with the aim is for shortening the development time and reducing errors in determining the product specification satisfying costumer needs [14]. This method is an effective method for new product development because it can identify customer needs and translate them into product attributes [15]. The AD and QFD approaches can be combined to develop market-competitive products. In the QFD analysis, provide data to designers about competitors and market opportunities so that they can provide a strong background for developing solutions. The advantages of this method are saving the product cost and increasing the market expectations [16].

2.5. Develop of Conceptual Design
The following are the steps of designing a stroller using the Axiomatic House of Quality (AHOQ) method [17]:
1. Conducting Benchmarking with the competitor products.
2. Develop the functional requirements (FR) to satisfy the customer attribute.
3. Transforming the functional requirements (FR) into the design parameters (DP)
4. Subsequently, constructing the AHOQ model and evaluating the AHOQ model that has been made.
5. Developing the concepts of product design based on the predetermined target specifications in the integration model between axiomatic design and house of quality.

2.6. Statistical Analysis
Non-parametric statistical analysis was conducted to test the hypothesis. They are dimension of stroller developed meet user requirements. It implemented the homogeneity method [18]. Moreover, design of stroller developed was significance difference with the existing design. It applied the Wilcoxon method [19]. The level of significance in these test is 5%.

3. Results and Discussion

3.1. Costumer Attributes Analysis
Table 1 presents the results of the customer survey that is the valid attributes required by user. Simple attribute shows that the design developed can accommodate user need especially it can reduce complications such as complexity in packaging, maneuvering, the process of opening and closing the canopy, and arrangement the baby equipment. As for a comfortable attribute is meant that user required no pain in use and likewise with safe attribute, the stroller can reduce the risk of accidents on user.

Table 1. Result of the Costumer Survey

Attribute	*Score*
Simple	72,8%
Comfortable	66,3%
Safe	56,4%
Durable	51,5%
Easy to use	48,9%
Serviceability	30,4%
Aesthetic	38%
Affordable Price	7,7%
Adjustable	17,5%

Next attribute that user require is strong and durable in which the stroller can withstand the load as well as may be used for a long time. Then attribute of easy to use is important for the user to operate and to maneuver and to climb when no difficulties in use. User also need the developed stroller can be repaired easily (serviceability). On the other hand, the aesthetic attributes are required for user to provide a self- satisfaction. Another important attribute is affordable price which it may be consideration for user to buy. The last attribute is adjustable that is user wants the stroller must be easy to adjust in operations.

3.2. Benchmarking Analysis
Benchmarking is conducted with making a comparison of the existing several brands of stroller to get some information to help in an effort to design the innovating stroller developed [20]. Table 2 shows the result of this benchmark.

Table 2. Result of Benchmarking Analysis

Attribute	***Importance Rating***	**The average value of *Cocolatte Pockit CL 83***	**The average value of *Babyelle Astro S350***
Simple	4	3.967	**4.167**
Comfortable	5	**3.967**	3.733
Safe	5	**4.233**	4.100
Multifunction	4	**3.867**	3.767
Durable	4	**3.633**	3.367
Easy to use	5	**4.233**	4.1
Serviceability	4	3.133	**3.967**
Aesthetic	4	**3.767**	3.667
Affordable Price	4	3.100	**3.833**
Adjustable	4	4.167	**4.200**

Based on the results of the benchmarking questionnaire above, the average value for each attribute in bold will be used as a reference in product design.

3.3. Result of Conceptual Design
The mapping process was conducted to determine Design Parameter of the stroller developed to satisfy the customer attribute based on the functional requirements.

Table 3. The Result of the Mapping Process Simple Attribute

Costumer Attribute	***Code***	**Functional Requirement**	**Code**	**Design Parameter**
Simple	**FR_1**	**Reduces complexity**	**DP_1**	**Minimalist design**
	$FR_{1.1}$	**Easy to package**	**$DP_{1.1}$**	**A frame design that does not have many connections**
	$FR_{1.1.1}$	Speed up the folding process	$DP_{1.1.1}$	The frame joints are equipped with a spring
	$FR_{1.2}$	Easy to manoeuvre	$DP_{1.2}$	A small amount of wheels
	$FR_{1.2.1}$	Easy when turning	$DP_{1.2.1}$	There are 3 wheels
	$FR_{1.4}$	Easy in the arrangement of luggage	$DP_{1.4}$	The basket is equipped with a bulkhead made of rubber

Table 3 shows the result of mapping process between design parameters and functional requirements to meet the simple attribute. The design of stroller should be minimalist for reducing the complexity in use. It is manufactured by installing a spring (DP 1.1.1) **cc.** Figure 4, in every joints of frame in which this spring can increase the speed and easy to folding. In other that, the stroller is set by making use of 3 wheels (DP 1.2.1) with 16,5 cm of diameter. A function of these wheels is in addition to easier in moving and also manoeuvring. Then to shade a baby it is required a retractable canopy (DP 1.3) that easy to open and close. The basket with a bulkhead is also installed in the frame to make easy in arranging some equipment (DP 1.4). For more detail see Figure 3.

Table 4 presents result of mapping process to meet the comfortable attribute in which it determines design parameter to satisfy the functional requirement of stroller developed. Ergonomic design (DP 2) is implemented to be design parameter of the stroller for reducing a negative impact of the body such as pain. Therefore, backrest with for 56.65cm in height and 29.4cm in width (DP 2.1.1.1) is required to support the back of baby and the seat with length 22.697cm (DP 2.2.1.1) is needed to reduce a pain in the back. The footrest is provided with 17.844cm (DP 2.2.2.1) of distance from seat in order to avoid pressure on the back of leg (DP 2.2.2.1). As for the thickness of seat cushion used is much 3cm (DP 2.2.3.1) to sustain the child's body weight such that it can eliminate stress on the buttock.

Table 4. The Result of the Mapping Process Comfortable Attribute

Costumer Attribute	*Code*	**Functional Requirement**	**Code**	**Design Parameter**
	FR_2	**Reduces pain in the body**	**DP_2**	**Ergonomic design**
	$FR_{2.1}$	Reduces pain in the upper body	$DP_{2.1}$	The design of chair backrest is according to the dimensions of the user's body
	$FR_{2.1.1}$	Reduces pressure on the back	$DP_{2.1.1}$	The size as the chair backrest
	$FR_{2.1.1.1}$	Allowed for a big child to sit freely	$DP_{2.1.1.1}$	95% percentile for seat height 56.65 cm, and seat width 29.4 cm
	$FR_{2.2}$	Reduces pain in the buttocks	$DP_{2.2}$	The design of the cushion is according to the dimensions of the user's body
	$FR_{2.2.1}$	Reduces pressure on the coccyx	$DP_{2.2.1}$	The length of the cushion
	$FR_{2.2.1.1}$	Provides flexibility for the average child of 6 months-3 years old in a not too long period of time	$DP_{2.2.1.1}$	50% percentile for seat length of 22,697 cm
	$FR_{2.2.2}$	Reduces tension on the back of the knee	$DP_{2.2.2}$	The height of the footrest is according to the the dimensions of the popliteal length
	$FR_{2.2.2.1}$	Allowed for a small child to reach the *footrest*	$DP_{2.2.2.1}$	5% percentile for footrest height of 17,844 cm
Comfortable	$FR_{2.2.3}$	Provides support for the child's weight	$FR_{2.2.3}$	Thickness of the foam for the cushion
	$FR_{2.2.3.1}$	Reduces pressure due to child's weight	$FR_{2.2.3.1}$	Thickness of cushion is 3 cm
	$FR_{2.3}$	Reduces pain in the palm	$DP_{2.3}$	The design of the *stroller* push is according to the dimensions of user's body.
	$FR_{2.3.1}$	Reduces the pressure due to the weight of the *stroller*	$DP_{2.3.1}$	The thickness of the sponge for the stroller push coating is 1 cm
	$FR_{2.3.2}$	Reduces pain in the wrist	$DP_{2.3.2}$	The diameter of the *stroller* push *handle* is according to the dimension of the hand grip
	$FR_{2.3.2.1}$	Increases grip strength for the smallest users	$DP_{2.3.2.1}$	5% percentile for the *stroller handle* push diameter of 5.101 cm
	$FR_{2.3.3}$	Prevents slip when hands are sweating	$DP_{2.3.3}$	Handlebar foam is equipped with rubber with the user's finger grooves
	$FR_{2.3.3.1}$	Improves grip strength for the average user	$DP_{2.3.3.1}$	50% percentile for rubber width (2,633 cm)
	$FR_{2.4}$	Minimizes the heat temperature on the cushion	$DP_{2.4}$	Cushion upholstery material
	$FR_{2.4.1}$	Easy to absorb the heat	$DP_{2.4.1}$	Parachute material

In other that, the stroller's handle is coated by 1 cm thick of sponge (DP2.3.1), 5.101cm (DP 2.3.2.1) of handlebar diameter, and 2,633cm width of finger grooves (DP 2.3.3.1) to prevent fingers slipped. The seat also coating with parachute material (DP 2.4.1) in order to absorbing the heat from the baby's body.

This following in Table 5 is the design parameter identified for meeting the safe attribute on basis the functional requirement that is safety design of the stroller for reducing the risk of accidents. Thus the design should be equipped by the pentagon safety harness system (DP 3.1.1) to maintain the baby body's position and, shock absorber on wheels (DP3.1.2) to decrease tremor when moving. While hand brakes (DP 3.2.1) is to easy in reach. And the locking pedals installed on the right, the left and the rear of wheel (DP 3.3.1) is to keep the position of stroller in stable, and uses a phosphor sticker on the stroller wheel (DP 3.4) is to clarify vision.

Table 5. The Result of the Mapping Process Safe Attributes

Costumer Attribute	*Code*	**Functional Requirement**	**Code**	**Design Parameter**
Safe	**FR_3**	**Reduces the risk of accidents**	**DP_3**	***Safety design***
	$FR_{3.1}$	Keeps children from falling from the *stroller*	$DP_{3.1}$	Safety system design
	$FR_{3.1.1}$	Maintains the child's body position	$DP_{3.1.1}$	5 safety harness system (5 side seat belts)
	$FR_{3.1.2}$	Able to absorb shocks	$DP_{3.1.2}$	shock absorber on wheels
	$FR_{3.2}$	Speed up reflexes when braking	$DP_{3.2}$	Braking system
	$FR_{3.2.1}$	The position of the brake is easy to reach when reflex	$DP_{3.2.1}$	Brakes are in hand
	$FR_{3.3}$	Keep the *stroller* wheel position stable when it stops	$DP_{3.3}$	Wheel locking system
	$FR_{3.3.1}$	Speed up the wheel locking process	$DP_{3.3.1}$	There are locking pedals on the right and left rear wheel connecting frame
	$FR_{3.4}$	Improves vision in dark places (glow in the dark)	$DP_{3.4}$	Uses a phosphor sticker on the *stroller* wheel

Table 6. The Result of the Mapping Process Durable Attribute

Costumer Attribute	*Code*	**Functional Requirement**	**Code**	**Design Parameter**
Safe	**FR_4**	**Increases product durability**	**DP_4**	**Strong Frame**
	$FR_{4.1}$	Increases resistance to chemical reactions	$DP_{4.1}$	Rust resistant material
	$FR_{4.1.1}$	Protects the frame against corrosion	$DP_{4.1.1}$	Aluminum alloy material
	$FR_{4.2}$	Able to support the weight of children until age of 3 years	$DP_{4.2}$	Maximum load of 35 kg

Table 6 presents the design parameter is generated from each functional requirement to satisfy durable attribute. In order to prevent corrosion and be able to withstand the burden of a 3-year-old child the stroller's frame is made from aluminium alloy material (DP 4.1.1) with 35kg maximum load (DP 4.2), as shown in Figure 5.

Table 7. The Result of the Mapping Process Easy to use and Serviceability Attributes

Costumer Attribute	*Code*	**Functional Requirement**	**Code**	**Design Parameter**
Easy to Use	**FR_5**	**Easy in using the *stroller***	**DP_5**	**Usable design**
	$FR_{5.1}$	Improves wheel stability	$DP_{5.1}$	Wheel diameter and thickness
	$FR_{5.1.1}$	Easy to move on uneven roads	$DP_{5.1.1}$	d = 16.5 cm, wheel thickness = 4cm
	$FR_{5.2}$	Improves wheel performance	$FR_{5.2}$	Design of wheel
	$FR_{5.2.1}$	Easy to use on roads that have stairs	$DP_{5.2.1}$	Installed with chain wheels
	$FR_{5.3}$	Increases user mobility	$DP_{5.3}$	*Lightweight design*
	$FR_{5.3.1}$	Easy to carry	$DP_{5.3.1}$	Maximum weight is 7 kg
Serviceability	**FR_6**	**Easy to repair**	**DP_6**	**Repair service**
	$FR_{6.1}$	Repair places are easy to find	$DP_{6.1}$	Can be repaired at a bicycle repair shop

As for Table 7 shows the result of the mapping process between design parameters and functional requirements to fulfil easy to use and serviceability attributes. Users need a stroller that is easy to carry, able to move on roads that have stairs, and stable wheels so it is provided the wheels with 16,5 cm in diameter, 4cm (DP 5.1.1) in thickness, Installed with chain wheels (DP 5.2.1), 7kg of maximum weight (DP 5.3.1). These dimension is intended for easy to move on uneven road as well as to improve stability and mobility along light to carry. Moreover, the stroller can be repaired easily.

Table 8. The Result of the Mapping Process Aesthetic and Affordable Price Attributes

Costumer Attribute	*Code*	**Functional Requirement**	**Code**	**Design Parameter**
Aesthetic	**FR_7**	**Provides an attractive product appearance for users**	**DP_7**	**Design based on *voice of customer***
	$FR_{7.1}$	Provides a modern impression on the product	$DP_{7.1}$	The color and pattern of the *stroller* cushion
	$FR_{7.1.1}$	Provides a futuristic impression	$DP_{7.1.1}$	Geometric pattern on the *stroller* seat
	$FR_{7.1.2}$	Provides an elegant impression	$DP_{7.1.2}$	RGB white color (255,255,255); RGB black color

Costumer Attribute	*Code*	**Functional Requirement**	**Code**	**Design Parameter**
				(0,0,0); RGB light gray color (211,211,211)
Affordable Price	**FR8**	**Able to reach most people (FR9)**	**DP8**	**Price IDR 1,750,000.00 - IDR 2,250,000.00**

While Table 8 presents the design parameter of attribute aesthetic and affordable price. It is in order to more attractive design and more futuristic with adding geometric pattern on seat (DP 7.1.1) as well as an elegant impression with white color (RGB: 255,255,255) and black color (RGB: 0,0,0) and light gray color (RGB: 211,211,211) (DP 7.1.2). This stroller price ranged from IDR 1,750,000.00 - IDR 2,250,000.00 (DP 8) so that the stroller can be reached by most people.

Table 9. The Result of the Mapping Adjustable Attribute

Costumer Attribute	*Code*	**Functional Requirement**	**Code**	**Design Parameter**
Adjustable	**FR9**	**Easy in changing the size of the product according to users' needs**	**DP10**	**Adjustable design**
	FR9.1	Speed up the process of changing the size of the push handle *stroller*	DP10.1	The push *handle* is equipped a joint frame.
	FR9.1.1	Reduces tension in the user's elbows, whether they have small, medium or large postures	DP10.1.1	High percentile impulse 5% 92,646 cm; 50% percentile 98,787; and 95% percentile 104,927 cm
	FR9.2	Provides ease of changing the shape of the *stroller* seat	DP10.2	The *stroller* seat is zipped to lay down the *stroller*

The last user requirement is adjustable. It make the stroller easier to change the product size according to user needs. Table 9 shows the stroller's handlebar can accommodate 3 size with percentile 5% is 92,646cm, 50% is 98,787cm, and 95% is 104, 927. The 3 types of the height of the handlebar are to reduce the tension of the user's elbow (DP 10.1). The stroller has a zipper to lay down the stroller's seat (DP 10.2).

For more details it can be seen in Figure 2 and Figure 3. They are about model of Axiomatic House of Quality and also the conceptual design of the developed new stroller.

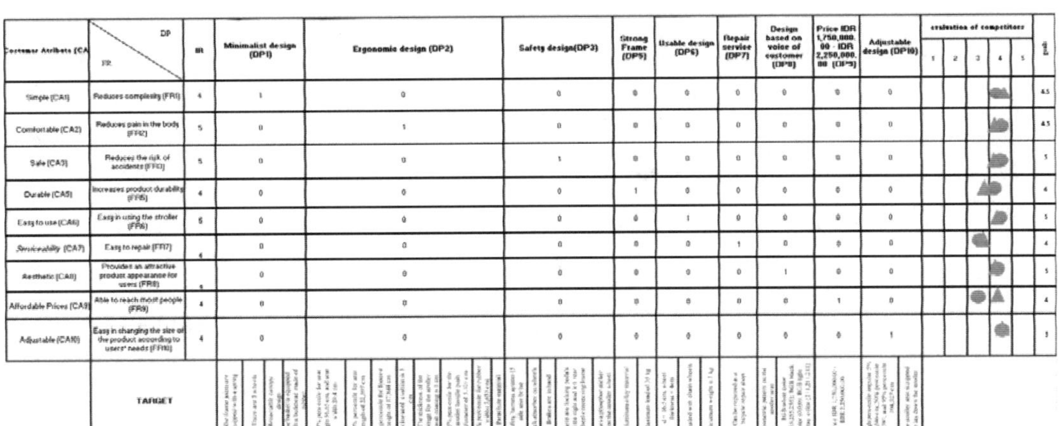

Figure 2. AHOQ Model

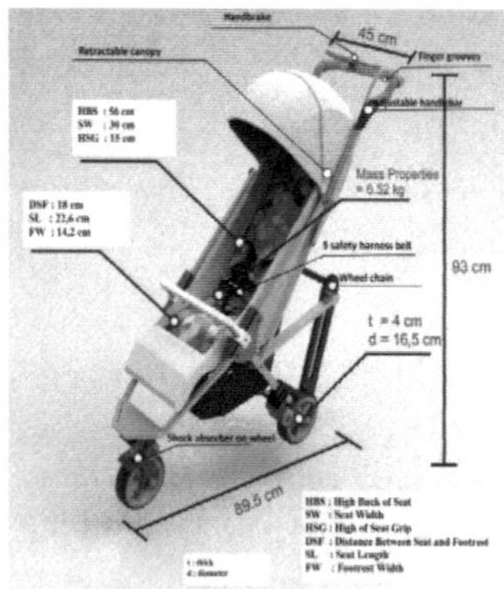

Figure 3. Conceptual Design of the Developed Stroller

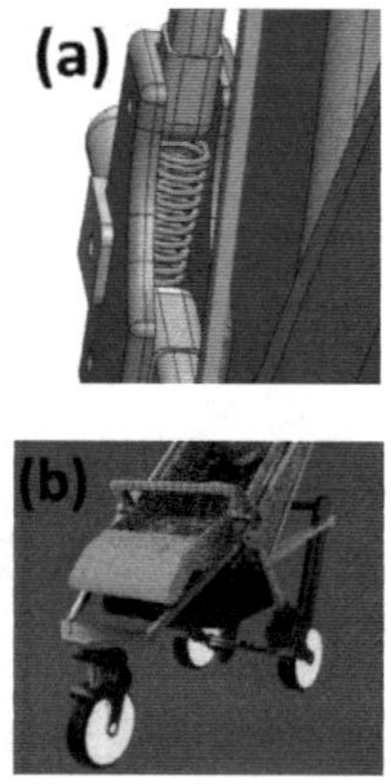

Figure 4. a) Spring at Frame Joint, b) Glow in the Dark Wheel

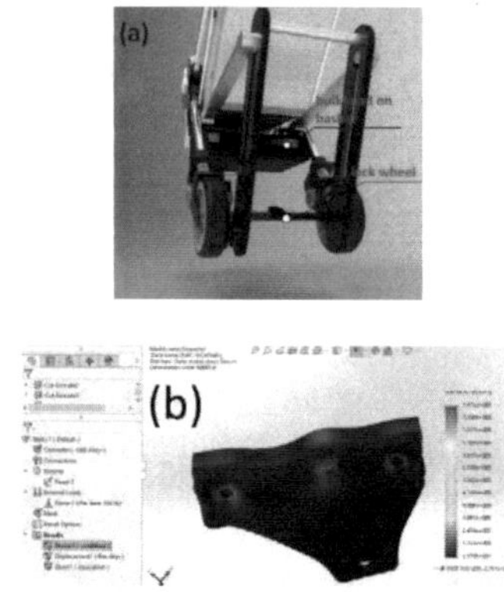

Figure 5. a) The Back of the Stroller b) Result of Load Test (maximum Load 35kg)

3.4. Validation Analysis

The Homogeneity test results from the designs proposed for validation presented in Table 10 at 5% of significance level. This test is characterized by hypotheses: there is no significant difference between the proposed design and the customer attribute (H_0). Results of this test is that H_0 is accepted. This fact shows the design of the stroller developed can meet the requirements of user.

Table 10. Result of Homogenity Test

Attribute	*Sig.Value*
Simple	0,150
Comfortable	0,128
Safe	0,182
Durable	0,057
Easy to use	0,057
Serviceability	0,862
Aesthetic	0,194
Affordable / cheap	0,075
Adjustable	0,109

Table 11. Result of Wilcoxon Test.

Attribute	*Sig.Value*
Simple	0,041
Comfortable	0,036
Safe	0,000
Durable	0,000
Easy to use	0,001
Serviceability	0,015
Aesthetic	0,000
Affordable / cheap	0,001
Adjustable	0,000

Table 11 presents the results of the Wilcoxon different tests at 5% of significance level. This statistical test is characterized by a hypothesis: there is no significant difference between the proposed stroller design and the stroller design before repairs (H_0). The test result in that null hypothesis is rejected. It means there is a significant difference between the proposed stroller design and the the existing design (H_1). Thus the proposed design developed is valid to more innovative and more ergonomic.

4. Conclusion

The results of the study concluded that,

a Customer Attributes from the developed stroller are simple, comfortable, safe, durable, easy to use, serviceability, aesthetically pleasing, affordable, and adjustable.

b Design parameters that are obtained from each attribute on the proposed stroller design are minimalist design, ergonomic design, safety design, usable design, and availability of repair services, designs based on the voice of the customer, prices IDR 1,750,000.00 - 2,250,000.00, and adjustable stroller design.

c The proposed stroller design is a valid to satisfy the costume requirements and valid to be innovative and ergonomics design at 5% of significance level.

Acknowledgement
The authors wish to acknowledge the Directorate of Student Development of Indonesian Islamic University in Yogyakarta-Indonesia for funding this project.

References
[1] Garret M, McElroy A, and Staines A, 2002, *Locomotor Milestone and Babywalkers*:
[2] Koh B X, Nirmal U, and Yuhazri M Y, 2019, *Developments on Baby Strollers over the Last Decade*. Current Journal of Applied Science and Technology, **33(6)**,1-25, 47713
[3] Sehat A R, Nirmal U, 2017, *State of The Art Baby Strollers: Design Review and The Innovations of an Ergonomic Baby Stroller*, Cogent Engineering, **4:1**, 1333273
[4] Wu X, Hong Z, Li Y, Zhou F, Niu Y, and Xue C, 2020, *A Function Combined Baby Stroller Design method Developed by Fusing Kano, QFD and FAST Methodologies*, International Journal of Industrial Ergonomics, **75**, 102867
[5] Hermantara J, Hariyano A, Halim A, Jesica, and Gunawan R, 2012, *Perancangan dan Pengembangan Stair Climb Stroller*, Program Kreativitas Mahasiswa - Karsa Cipta, 1-2.
[6] Utami P, and Indah R, 2012, *Perancangan Kereta Dorong Bayi Yang Lebih Ergonomis (Studi Kasus di Perusahaan Baby Champion)*, Laporan Tugas Akhir Universitas Maranatha, 11.
[7] Arthaya B, and Susanty, 2008, *The Redesign of Baby Carriage for Accomodating Stroller Car Seat of Baby Does, Chicco, and Pliko Types*, Proceedings of 9th Asia Pacific Industrial Engineering & Management System Conference (APIEMS 2008), Bali
[8] Pratama A, Sufyan A, and Adiluhung H, 2019, *Pengembangan Perancangan Stroller Dengan Meningkatkan Kenyamanan dan Fungsional Anak Usia 1 – 3 Tahun*, e-Proceeding of Art and Design: 6(1)
[9] Ahnert S, Dean R L, and Dotsey M A, 2014, *Foldable, Removable, Reversible Stroller Seat*, United States Patent, Patent No.: US 8,764.048 B1
[10] Du Z, Fan Y, Sullivan C, and Wen Y, 2010, *Safety Stroller, ME450*, FINAL REPORT Department of Mechanical Engineering University of Michigan, Ann Arbor, MI 48109-2125
[11] Suh N P, 2001, *Axiomatic Design–Advance and Application.* New York: Oxford University Press
[12] Suh N P, 2005, *Complexity: Theory and Applications*, New York: Oxford University Press.
[13] Lee D G, and Suh N P, 2006, *Axiomatic Design and Fabrication of Composite Structures: Application in Robots, Machine Tools, and Automobiles*, New York: Oxford University Press.
[14] Ashtiany M, and Alipour A, 2016, *Integration Axiomatic Design with Quality Function Deployment and Sustainable Deisgn For Satisfaction of an Ariplane Tail Stakholders*, Procedia CIRP.
[15] Chaucick P M, Carnevalli J A, and Calarge F A, 2007, *Using Axiomatic Design for Minimizing QFD Application Difficulties in NDP: Research Proposal and Preliminary Definition of First and Second Hierarchical Levels*, Product: Management Development, 127-132.
[16] Taglia A, and Campatelli G, 2006, *Axiomatic Design QFD: A Case Study of a Reverse Engineering System for Cutting Tools*, Proc. 4th Int. Conf. Axiomat.Des. (ICAD06).
[17] Machulenko N, 2001, *Appliying Axiomatic Design Principles to House of Quality*, Thesis Ontario: University of Windsor.
[18] Yamin S, and Kurniawan H, 2009, *SPSS Complete*, Jakarta: Salemba Infotek.
[19] Pramana A, and Mawardi W, 2012, *Analisis Perbandingan Trading Volume Activity dan Abnormal return Saham Sebelum dan Sesudah Pemecahan Saham*, Diponegoro Journal Management, **1**, 5.

Energy harvesting applications using 3D-printed coreless generator

Safarudin Gazali Herawan[1], Said Abi Syahputra[1], Ernie Mat Tokit[2], Fatimah Al-Zahra Mohd Sa'at[2], Mohamad Afzanizam Mohd Rosli[2]

[1]Industrial Engineering Department, Faculty of Engineering, Bina Nusantara University, Jakarta, Indonesia 11480

[2]Fakulti Kejuruteraan Mekanikal, Universiti Teknikal Malaysia Melaka, Hang Tuah Jaya, 76100 Durian Tunggal, Melaka, Malaysia.

Email: safarudin.gazali@binus.edu

Abstract. 3D-printed coreless generator has been presented for energy harvesting applications using permanent magnet as source of external magnetic field in the rotor. The phototype of proposed coreless generator is built using 3D-printer with Black Polylactic Acid (PLA) as a filament for stator housing, rotor and other parts. The major parts of the proposed coreless generator are the stator, which has four coils of winding copper wire and the rotor with eight permanent magnets which are arranging N and S magnet poles alternately. This research is conducted to determine how much power can be generated from the proposed 3D-printed permanent magnet generator. The prototype permanent magnet generator is tested with rotational speed and load condition. The experimental results have been shown that the coreless generator can generate the electrical energy up to 5 W at 6000 rpm.

1. Introduction

Permanent magnets generators (PMGs) exhibit a prospect to build energy efficient machine [1]. The stator and rotor are the major parts of PMGs. The main function of the stator is to generate alternating current from conductor windings while the rotor consists of permanent magnets as external magnetic field to generate the magnetic interaction with the stator. Nowadays PMGs are rapidly used in numerous applications in term of powering small or large-scale electronic systems, such as wind turbine application [2], hydroelectric generation [3] and others. Even though PMGs are able to achieve high efficiency compared with other generator types, however PMGs still more costly compared to other generators [4]. Therefore, many research works focus on the design of PMGs with basic requirements such as high output power, lightweight, simple construction, and low cost [1]. Unfortunately, the design consideration of those basic requirements should be carefully conducted based on the generator configuration and materials [4].

3D printing technology has become significant importance as rapid process to develop prototypes for various type of applications such as RF/microwave components [5], aerial vehicle wing structures [6], electromagnetic shielding [7] and biomedical applications [8]. Since the 3D printing technology has capabilities to develop complex geometries from the digital drawing, this technology becomes the major advantage in the manufacturing systems. The other advantage is that the 3D printing technology can be used to develop functionalities with other manufacturing methods [9]. Additionally, 3D printing technology can improve the design and lower the weight of the structure, resulting to a very cost-effective solution and reduction of lead times [10]. Based on these results, it is a worthy probability to use 3D printing technology for developing a generator. The development of generator using 3D-printing technology have been proposed by many researches works such as 3D-printed electrostatic generator [11], 3D-printed triboelectric generator [12], and 3D-printed materials low speed PM generator [13].

In this work, the possibilities of implementing 3D printed coreless generator for energy harvesting applications is reported. Following sections will explain the design of the 3D-printed coreless generator, fabrication of the generator, and the test results of the proposed 3D-printed coreless generator.

Content from this work may be used under the terms of the Creative Commons Attribution 3.0 licence. Any further distribution of this work must maintain attribution to the author(s) and the title of the work, journal citation and DOI.
Published under licence by IOP Publishing Ltd

2. Materials and methods

2.1. Design of 3D printed coreless generator

The generator is meant to convert mechanical energy into electrical energy. However, the coreless generator is a generator that use the core for coil of winding copper wire by non-ferrous materials. Most of these type of generators used Permanent Magnet (PM) in the rotor. PM in generator supplied the magnetic excitation as conversion device producing all electromagnetic energy. The advantage of coreless generator is that it can eliminate the strong interaction in magnetic field between PM and ferrous core resulting to a high resistant on rotating rotor. Therefore, the coreless generators are suitable for harnessing renewable energy such as wind, rain water, river water, and tidal energy, which produce relatively small energy. Hence, the coreless generator can capture a small amount of energy by smooth rotation without any resistant of magnetic field.

The main parts of the generator are stator and rotor. There are two types of generator; the first type has stator which is fixed at the middle and a rotor that is located at the outer stator housing, and the other type has rotor at the middle and the stator is fixed at the outer housing. The stator functioned as conductor winding to generate alternative current. While, the rotor, where the PM took place, is operating as external magnetic field to generate the magnetic interaction with the stator, thus can produce the alternative current.

All parts of the coreless generator are designed using Autodesk Inventor. The design of stator and rotor are as shown in figure 1 and figure 2, respectively. In the stator design, there are four stacks for winding coil of copper wire that are placed perpendicular to the next coil. The rotor as shown in figure 2 is designed for 8 permanent magnets in the slot for easier placement. The stator is the stationary part in the outer housing whereas rotor is the inner part which rotates. The specification of designed coreless generator are summarised in table 1.

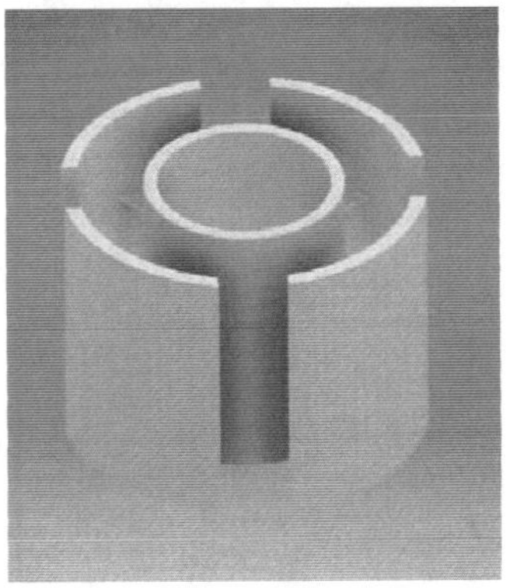

Figure 1. Stator design in Inventor Autodesk

Figure 2. Rotor design in Inventor Autodesk

Table 1. 3D printed coreless generator specification

Parameters	Value
Pole numbers	8 poles
Stator Slot for winding	4 slots
Rotor diameter	45 mm
Length of stator stack	39 mm
Stator inner diameter	50 mm
Stator outer diameter	97 mm
Air gab length	2,5 mm
Magnet dimension	60 x 10 x 5 mm^3

2.2. 3D printed coreless generator fabrication

Anet A8 3D printer is used for the coreless generator fabrication with 1.75 mm of diameter of Black Polylactic Acid (PLA) as a 3D printing filament. Before running the 3D printer, the design of stator and rotor should converted to stl file, then the file is opened in Cura software for printer setting as shown in figure 3. The setting are; the heating bed temperature and nozzle temperature set to 60°C and 210°C, respectively. The printing speed is 60 mm/s with the density of 60%. The support placement was chosen as "Everywhere" type to support hanging profile and can be removed easily without effecting the whole product. Figure 4 shows the construction of stator during operation of 3D printer.

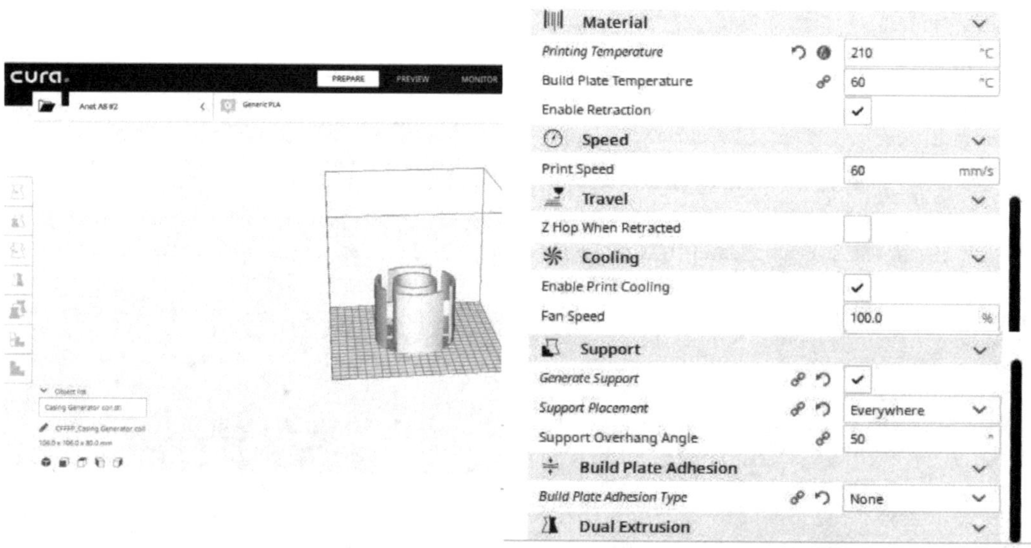

Figure 3. Configuration setting on Cura Software for 3D printing

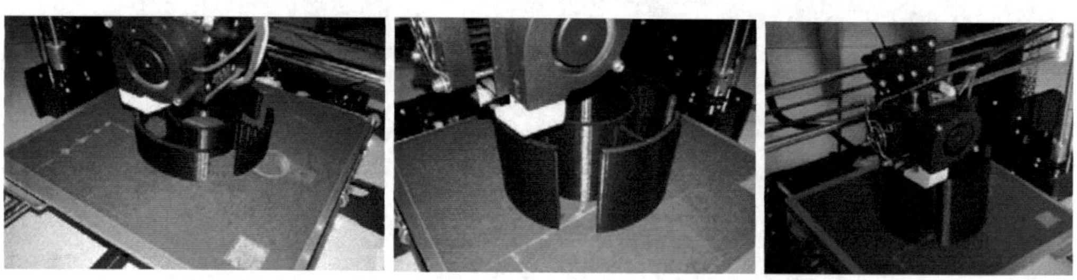

Figure 4. Coreless generator fabrication using Anet A8 3D printer

A 0.60 mm copper conductor wire is used for winding coil for four slots of the stator with 500 turns in each slot as presented in figure 5. It is essential to count the turns for winding correctly and carefully. One turn of copper wire is set next to the other with enough tense and less gaps in between the turns of copper wire [4]. This is done to produce successful coils. At the rotor, 8 poles neodymium magnets are chosen due to stronger magnets with less volume to produce the external magnetic fields as shown in figure 6. The arrangement of the N and S magnet poles are set alternately.

Figure 5. Stator with winding copper wire

Figure 6. Rotor with 8 poles permanent magnet and arrangement

2.3. Experimental setup

In conducting this prototype, the coreless generator was tested using load of 12 V bulb. The prototype of coreless generator can be seen in figure 6(a), while the diagram of experimental setup can be seen in figure 6(b). The connection wire between the coils and connection to a bridge rectifier to convert AC to DC current, load, voltmeter, and amp meter are clearly shown in figure 6(b). The prototype of the coreless generator was tested with rotational speed ranging from 0 rpm up to 6000 rpm.

The experiments were conducted with three different conditions. First, the coreless generator was running without any load to observe an amount of voltage that could be produced. Second, the coreless generator was running with 1 load of 12 V 5 W bulb. Lastly, the coreless generator was running with 2 loads of 12 V 5 W connected parallelly. For second and third experiments, the current produced from the coreless can be observed by Amp meter and Voltmeter for monitoring voltage. The coreless generator was connected to 100 W AC motor to drive the rotor during the experiments that can be controlled by speed controller from 0 rpm up to 6000 rpm.

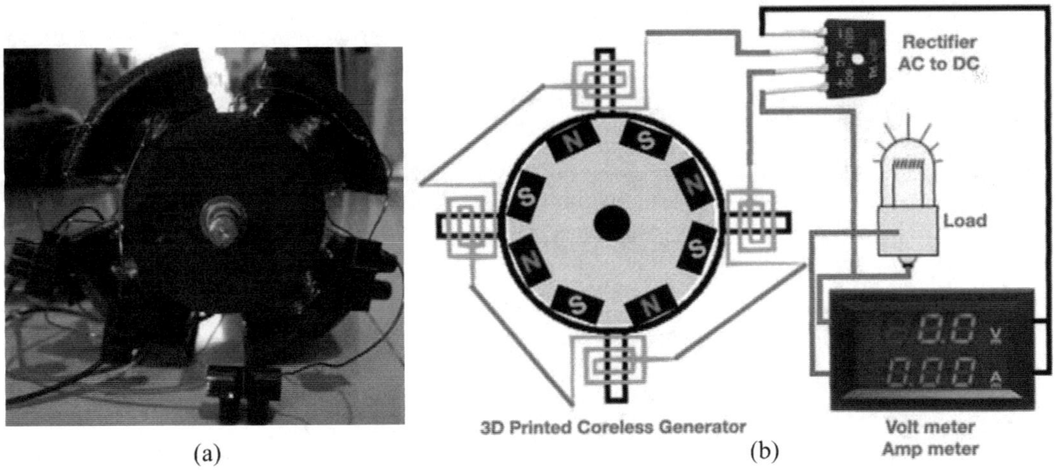

(a) (b)

Figure 6. 3D printed coreless generator (a) and diagram of experimental setup (b)

3. Results and Discussion

Figure 7 summarized the voltage produced from the coreless generator. It can be observed that the increasing voltage is depending on the increasing of speed linearly. From the experimental work, the proposed generator can generate voltage of up to almost 25 volt at 6000 rpm. Therefore, it can be concluded that this coreless generator needs higher speed to get higher voltage. Therefore, for efficient energy conversion related to the use of renewable sources, identification of the gear ratio between generator and turbine is crucial to obtain higher voltage.

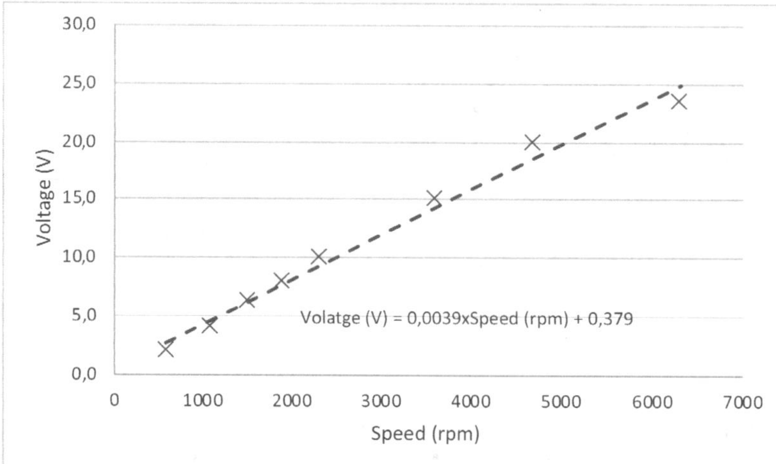

Figure 7. 3D printed coreless generator voltage with increasing speed

By applying loads at proposed generator during the experimental work, the production of voltage reduces. If higher load is implemented, at lower voltage will be generated as shown in figure 8. However, the increasing load can increase the output current. At 6000 rpm, the voltage without load can generate almost 25 volt, then by giving 1 load and 2 load, the voltage reduce to 17 volt and 10 volt, respectively. While the output current by giving 1 load and 2 loads, can increase from 350 mA to 500 mA at 6000 rpm.

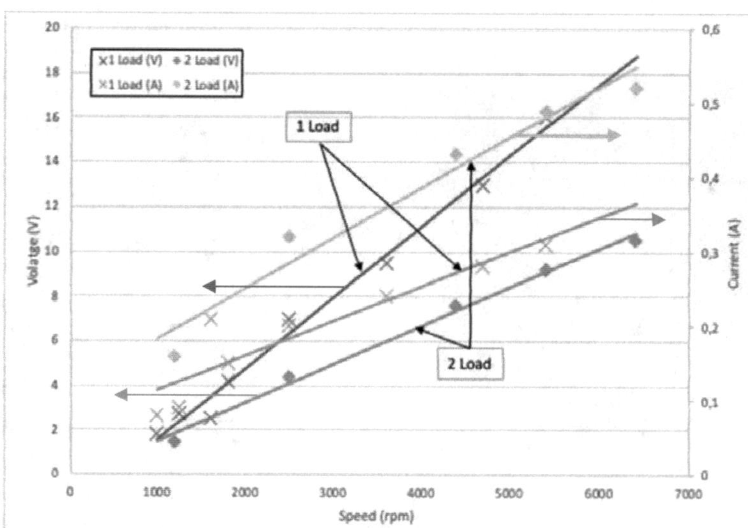

Figure 8. 3D printed coreless generator voltage and current with increasing speed at 1 and 2 Loads

Figure 9 shows the output power produced from the coreless generator with 1 load and 2 loads. It is clearly shown that both of loads can produce similar results, eventhought, the voltage and current produced from 1 load and 2 loads are different. This coreless generator can produce up to 5 W at 6000 rpm. Therefore, it can be concluded that regardless of loading applied on this coreless generator, the generator will produce in similar output power especially under the two loads conditions as tested in this study.

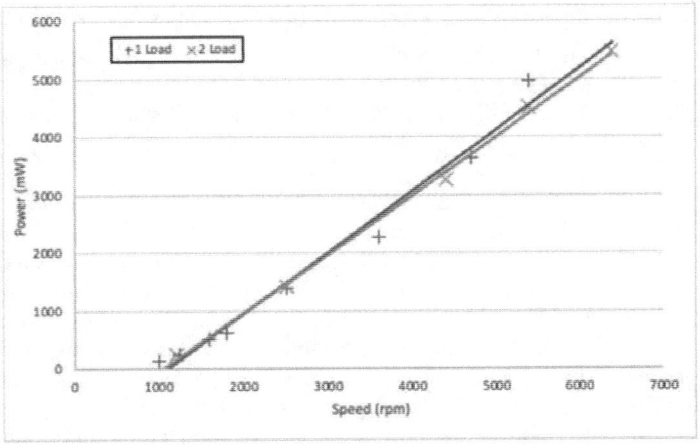

Figure 9. 3D printed coreless generator power with increasing speed at 1 and 2 Loads

4. Conclusions

In this work, the experiments with prototype of coreless generator created using 3D-printed are presented. The coreless generator is aimed to be applied in energy harvesting system. It was found that 3D-printed coreless generator can generate electric energy up to 5 W at 6000 rpm that can be used in low-power system. Therefore, the development of coreless generator using 3D-printed exhibits promising potential to be used for this generator. Nevertheless, further studies are required to study the process developments in variety of filaments and setting on 3D printer.

Acknowledgements

This work is supported by Research and Technology Transfer Office, Bina Nusantara University as a part of Bina Nusantara University's International Research Grant entitled Development of Siphonic-Roof-Top Water Turbine for Green Technology with contract number: No.026/VR.RTT/IV/2020 and contract date: 6 April 2020.

References

[1] Kallaste, Ants, Vaimann, Toomas, and Pabut, Ott. "Slow-speed ring-shape permanent magnet generator for wind applications." 11th International Symposium of Typical Problems in the Field of Electrical and Power Engineering (2012): 66-69.
[2] Soderlund L, Eriksson J T, Salonen J, Vihriala H and Perala R 1996 A permanent-magnet generator for wind power applications *IEEE Transactions on Magnetics* **32(4)** 2389-92
[3] Huzlik R, Ondrusek C and Palsovics N 2016 December. Design of permanent magnet generators for small hydro power plant *In 2016 17th International Conference on Mechatronics-Mechatronika (ME)* 1-5 IEEE
[4] Latoufis K C, Messinis G M, Kotsampopoulos P C and Hatziargyriou N D 2012 Axial flux permanent magnet generator design for low cost manufacturing of small wind turbines *Wind engineering* **36(4)** 411-31
[5] Garcia C R, Rumpf R C, Tsang H H and Barton J H 2013 Effects of extreme surface roughness on 3D printed horn antenna *Electronics letters* **49(12)** 734-6
[6] Moon S K, Tan Y E , Hwang J and Yoon Y J 2014 Application of 3D printing technology for designing light-weight unmanned aerial vehicle wing structures *International Journal of*

Precision Engineering and Manufacturing-Green Technology **1(3)** 223-8

[7] Viskadourakis Z, Vasilopoulos K C, Economou E N, Soukoulis C M and Kenanakis G 2017 Electromagnetic shielding effectiveness of 3D printed polymer composites *Applied Physics A*, **123(12)** 736

[8] Wang K, Wu C, Qian Z, Zhang C, Wang B and Vannan M A 2016 Dual-material 3D printed metamaterials with tunable mechanical properties for patient-specific tissue-mimicking phantoms *Additive Manufacturing* **12** 31-7

[9] Haque R I, Farine P A and Briand D 2017 3D printed materials based triboelectric device for energy harvesting and sensing *In Multidisciplinary Digital Publishing Institute Proceedings* **1(4)** 580

[10] Mueller B 2012 Additive manufacturing technologies–Rapid prototyping to direct digital manufacturing *Assembly Automation*

[11] de Queiroz A C M and de Oliveira Filho L C M 2016 Energy harvesting with 3D-printed electrostatic generators *In 2016 IEEE 7th Latin American Symposium on Circuits & Systems (LASCAS)* 127-30 IEEE

[12] Quan S, Bu L, Zhou Z, Xu W, Liang B and Song L 2016 3D-printed novel triboelectric generator based on saw-toothed button structure *In 2016 13th IEEE International Conference on Solid-State and Integrated Circuit Technology (ICSICT)* 263-5 IEEE

[13] Soemphol C and Angkawisittpan N 2020 3D-printed materials based low-speed permanent magnet generator for energy harvesting applications *Materials Today: Proceedings* **22** 180-4

IOP Conf. Series: Materials Science and Engineering 1082 (2021) 012005 doi:10.1088/1757-899X/1082/1/012005

Nutmeg oil production process analysis using Business Process Improvement-a case study

R Darma[1*], H Y Sastra[2,5], Arhami[3], F Erwan[4,5]
[1]Department of Industrial Engineering, Politeknik Aceh Selatan, Tapak Tuan, Aceh, Indonesia
[2]Industrial Engineering Postgraduate Department, Universitas Syiah Kuala, Darussalam, Banda Aceh, Indonesia
[3]Department of Mechanical and Industrial Engineering, Universitas Syiah Kuala, Darussalam, Banda Aceh, Indonesia
[4]Industrial Engineering Department, Universitas Syiah Kuala, Darussalam, Banda Aceh, Indonesia
[5]Systems Design and Industrial Management Laboratory, Industrial Engineering Department, Universitas Syiah Kuala, Darussalam, Banda Aceh, Indonesia

E-mail: friesca_erwan@unsyiah.ac.id

Abstract. This study delivers the analysis of business process of the nutmeg oil production process in a small and medium enterprise (SME) in South Aceh using Business Process Improvement (BPI). This study analyses the pre-refining, refining and post refining activities of the nutmeg oil production process based on economics, labour, and technology aspects. This study carries out 3 phases of BPI which are organizing for improvement, understanding the process, and streamlining. This study reveals activity efficiency from 31 (on the as-is model) to 26 (on the to-be-model). The activities are further analysed using Devil's Quadrangle, particularly for cost and time dimensions. As a result of the new technology implementation being proposed, the fixed cost has increased by 48.33%, the variable cost has decreased by 47.64% and time has improved by 33.33%. Moreover, this study provides a recommendation to implement the to-be-model of the business process as an SOP for the nutmeg oil production process in South Aceh SMEs.

1. Introduction

Small and Medium Industries (SMIs) are one of industrial development sectors which have major impact on employment and contribute to reduce poverty alleviation in Indonesia [1]. SMIs need to increase their competitiveness through innovation [2]. However, since the industry scales are small and medium, SMIs encounter challenges in relation to limited working capital, low quality of human resources, and lack of mastery in technology. The challenges occur in most SMIs particularly in developing regions in Indonesia such as South Aceh.

South Aceh is one of the nutmeg producers in Indonesia. The nutmeg plant is a superior commodity for the South Acehnese that contributes to economic regional growth. According to South Aceh District Agency of Trade, Industry Cooperatives and SMEs, there are 66 SMEs of nutmeg oil processing in 2018 [3]. The significant numbers of SMEs need to be maintained both for their sustainability and development thus support regional economic growth continuously.

Content from this work may be used under the terms of the Creative Commons Attribution 3.0 licence. Any further distribution of this work must maintain attribution to the author(s) and the title of the work, journal citation and DOI.
Published under licence by IOP Publishing Ltd

Domestic demand for nutmeg oil is high. However, the significant market prospectus does not necessarily create a stable condition for the nutmeg oil processing SMEs. Various factors such as limited access to investment, lack of human resources skills and appropriate supporting technology create obstacles in the SMEs development. To analyse the obstacles, this study pay more attention on the business process activities of nutmeg oil production process using Business Process Improvement (BPI). A case study at UD. Permata Bunda South Aceh was carried out to obtain relevant data needed in this study.

BPI leads to quality and service improvement, cost efficiency, and productivity improvement of business process activities [4]. The main of objectives of BPI are to solve existing challenges of particular business process activities [4] and improve ongoing business processes by providing recommendations to solve issues in the business process [5]. BPI owns five phases (Figure 1) in optimising a business process, which includes organizing for improvement, understanding the process, streamlining, measurement and controls, and continuous improvement. The use of BPI in this study will assist in systematic analysing of the nutmeg oil processing activities which includes existing business process activities (as-is model), activities which need to be improved, and proposed business process improvements (to be model). It is expected that the clear structure of BPI will contribute in the development of nutmeg oil processing SMEs in South Aceh.

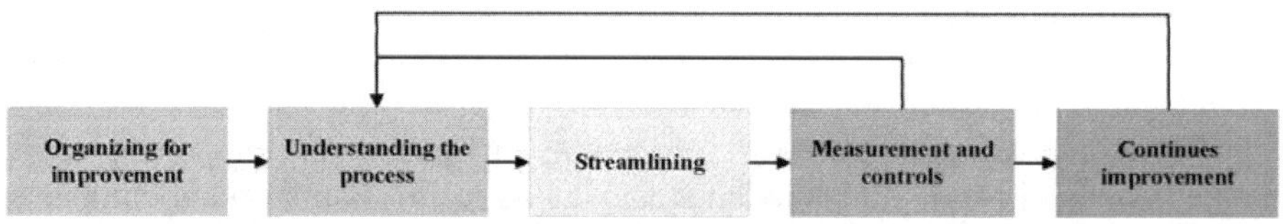

Figure 1. Business Process Improvement framework [4]

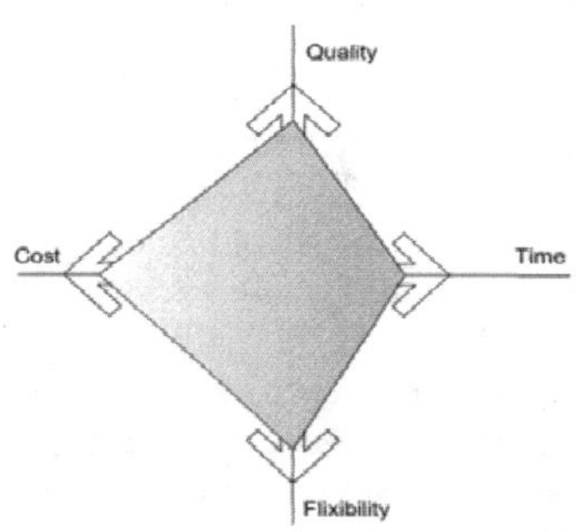

Figure 2. Devil's Quadrangle [6]

BPI is an improvement tool, however it needs to be measured. According to [4], significant improvements of BPI results can be described through comparing the before-and-after of time and cost of the business process being investigated. The improvement of cost and time will likely to improve other elements in the business process such as efficiency, effectiveness, customer satisfaction and Return on Investment (ROI) [4]. One of the most common tools to measure cost and time and describe its improvement is Devil's Quadrangle (Figure 2). The measurement tool was developed and introduced by [6] which consists of cost, quality, time, and flexibility. In their study, [6] explained that the redesigned model of business process ideally would improve time needed to handle orders, reduce costs of operational business processes, improve quality of services and business process capabilities.

Given the explanation of BPI and devil's quadrangle, this study aims to analyse the existing business process activities in the nutmeg oil processing by constructing the existing (as-is) and redesigned (to be) model of the business process activities using BPI phases, then measured and analysed the results using devil's quadrangle to provide recommendations for the development of nutmeg oil processing SMEs in South Aceh.

2. Research method

2.1. Data collection

This study occupies a case study method through sample selection process. Amongst the 66 nutmeg oil processing SMEs in South Aceh, only 13 SMEs conduct production process activities regularly [3]. Based on field observation, authors decided to select one SMEs as sample for the case study. To select the most supporting SME to deliver the case study, this study ranks the 13 nutmeg oil processing SMEs as shown in Table 1. The selection process was made according to technical requirements of the Directory of Micro, Small and Medium Enterprise of South Aceh District Agency of Trade, Industry Cooperatives and SMEs [3] and the similarity of production process activities amongst the SMEs.

Table 1. Nutmeg oil processing SMEs rank and selection results

SMEs	Manpower	Production capacity (kg)	Rank
UD. Permata Bunda	5	7200	1
UD. Syarifuddin	2	6000	2
Usaha Waldizar	2	5400	3
UD. Arafah Jaya	2	5000	4
UD. Zamzami	2	5000	5
Usaha Muhammad	2	2400	6
UD. Adek	2	2200	7
Intanmar	2	2000	8
UD. Palatama Raya	4	2000	9
Usaha Arsyad Aldy	2	1500	10
Usaha Hilmi.K	2	1500	11
UD. Titipan Kanda	4	950	12
Ketel Rustam	2	925	13

2.2. Data analysis

This study employs BPI phase 1 to 3. The business process being investigated is described in phase 1, the existing business process model (as-is model) is built on phase 2 and the redesigned model (to be model) is built on phase 3. The business process models, both for as-is and to be, are constructed following Business Process Modelling and Notation (BPMN) using a software named Bizagi modeller. Furthermore, the to be model further analysed using devil's quadrangle time and cost dimensions. For cost dimension, a simple economic measurement is carried out to easily compare cost improvement between as-is and to be model.

3. Result and discussion

3.1. Organizing for improvement

Organizing for improvement is the first phase of BPI which aims to review and select critical process within the company, and determine the business process of the critical process to be improved [4]. A case study which conducted at UD. Permata Bunda shows that this SME obtains raw material of the

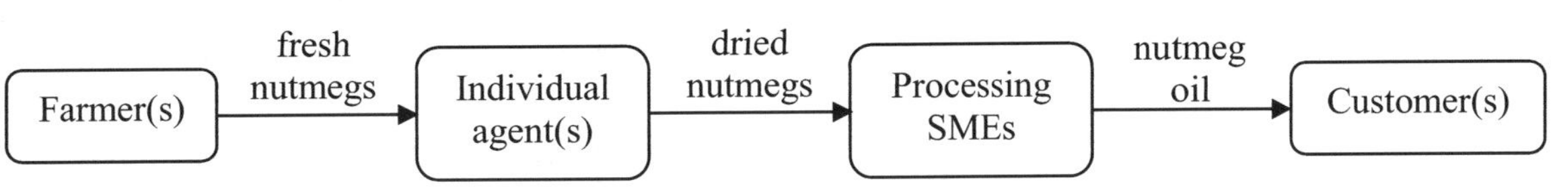

Figure 3. Nutmeg oil supply chain activities

nutmeg oil in a form of dried nutmegs from any of individual agents. The dried nutmegs will further processed to be nutmeg oil and distributed to customers. Figure 3 shows supply chain activities of nutmeg industry in general, which also applied in UD. Permata Bunda.

In a month, UD. Permata Bunda is able to conduct 4 times production activities. For a single nutmeg oil processing, this SME employs 5 temporary labours and utilises 4 boilers to refine 2400kg dried nutmeg into 240kg oil in 72 hours production time. The production processes consist of pre-refining process, refining process and post-refining process. UD. Permata Bunda is the top 5 active nutmeg oil processing SMEs in South Aceh. Based on Table 1, its production capacity reaches 7200kg. However, the production process still uses conventional technology. such as using wood instead of electricity or renewable energy. Despite of environmental issue and dependency of using wood, the process takes longer time and needs more labours to work. Table 2 shows the business process description of nutmeg oil production process in UD. Permata Bunda. The SME aims to improve the business process through substituting the conventional technology thus reducing production cost and time.

Table 2. Business process description of nutmeg oil production process

Process	Description
Business process definition	Nutmeg oil production process.
Objective	Deliver a refining process and produce a qualified nutmeg oil.
Internal environment	Production facilities.
External environment	Temporary labours. Individual agent(s) and customer(s).
Input	Dried nutmeg, wood (energy).
Output	Nutmeg oil.

3.2. Understanding the process

Understanding the process is the second phase of BPI which aims to understand the ongoing business processes by constructing current business process (as-is) model and analysing the existing activities contained in the model. Figure 4 shows nutmeg oil production process which divided into pre-refining, refining, and post refining process. The existing activities in the production process then transformed into a business process model following BPMN using Bizagi modeller (Figure 5). The simulation using Bizagi resulted that the business process activities consists of 31 main activities, 1 start event, 2 intermediate events, and 1 end event.

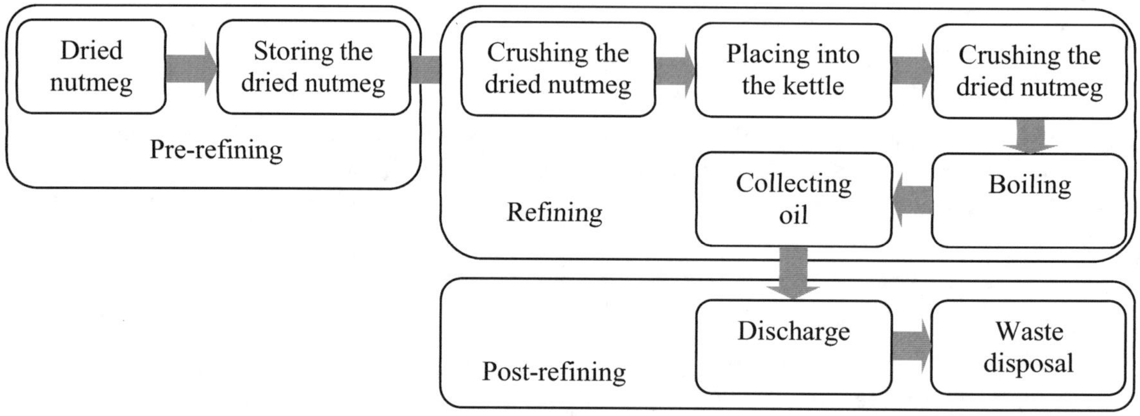

Figure 4. Nutmeg oil production process

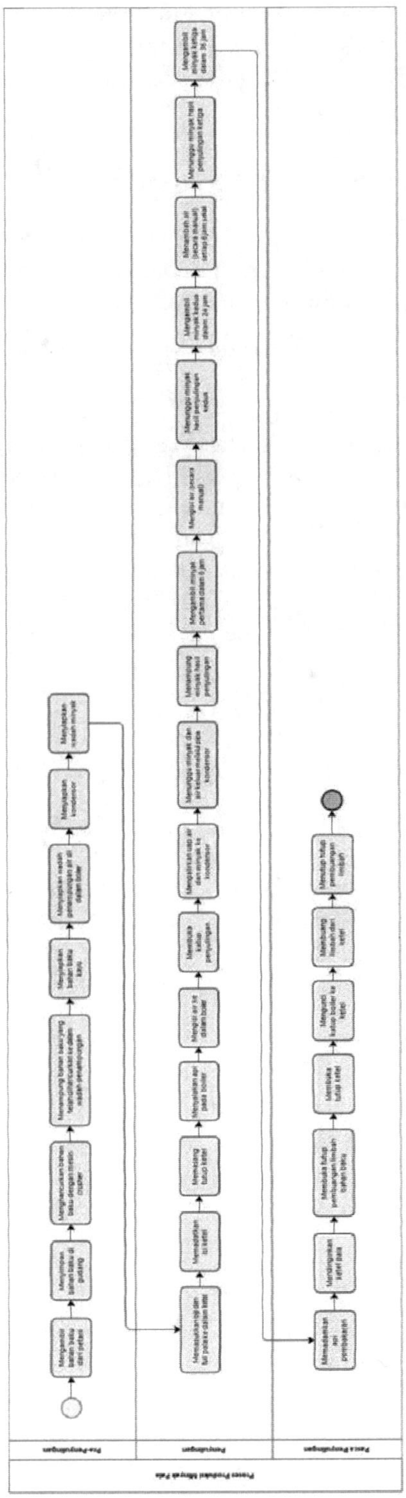

Figure 5. As-is model of nutmeg oil production activities

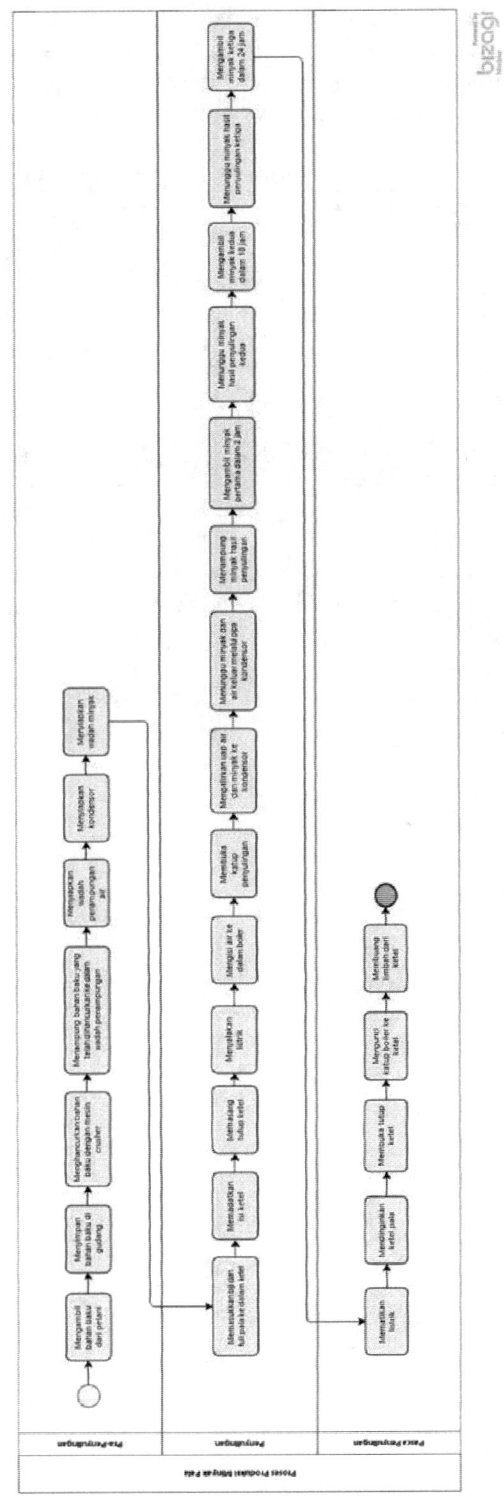

Figure 6. To be model of nutmeg oil production activities

3.3. Streamlining

The third stage of BPI is improving the business process through analysing the current business process. In this study, the improvement focuses on technology substitution in the refining process to reduce production cost and time. Table 3 shows the improving activities using streamlining tools and Figure 6 shows the to be model of the business process. The Bizagi modeller of the proposed model consists of 26 tasks, 1 start event, 2 intermediate events, and 1 end event.

Table 3. Improving activities using streamlining tools

Streamlining tools	Recommendations
Upgrading	Substitute conventional technology (using firewood) with hybrid technology (using electricity power and controller), thus reduce production time and eliminate manual activities.
Process cycle time reduction	Eliminate manual and unnecessary activities.

3.4. Devil's Quadrangle

To further analyse and measure the efficiency of substituting conventional technology (using firewood) with hybrid technology (using electricity power and controller), this study employs Devil's quadrangle analysis and measurement for cost and time dimension. For cost dimension, a simple economics analysis has been carried out. The cost components are divided into fixed and variable cost as shown in Table 4 and Table 5. Fixed cost contains item costs which will remain constant and not depend on number of goods or services produced, such as building rent. While variable cost contains item costs which will change depend on the production process, such as the volume of goods or products produced by the SMEs.

Table 4. Fixed cost analysis (in Rupiah)

No	Cost item	As-is	To be model
1	Production facilities		
	- Crushing machine	13.000.000	13.000.000
	- Boiler	72.000.000	222.952.000
	- Kettle	120.000.000	190.160.000
	- Condenser	13.000.000	13.000.000
	- Oil container	1.600.000	1.600.000
	- Controlling equipment	-	11.250.000
2	Rent	20.000.000	20.000.000
3	Electricity power	7.200.000	7.200.000
4	Water	1.800.000	1.800.000
5	Manpower insurance	362.880	362.880
6	Maintenance (1%)	2.196.000	4.519.620
7	Depreciation (1%)	2.196.000	4.519.620
	TOTAL	**253.354.880**	**490.364.120**

Table 5. Variable cost analysis (in Rupiah)

No	Cost item	As-is	To be model
1	Direct labour cost	67.200.000	67.200.000
2	Energy/fuel cost	144.000.000	20.509.056
3	Transportation (for raw material)	48.000.000	48.000.000
	TOTAL	**259.200.000**	**135.709.056**

Table 4 shows that fixed cost increase from Rp 253.354.880 to Rp 490.364.120 or equivalent with 48.33%. As a result of applying hybrid technology, SME need to employ applicable boiler and controller that suit the technology. However, this cost applied once for purchasing cost and will not affect the entire operational cost. In addition, Table 5 shows applying hybrid technology will reduce variable cost since fuel usage could be diminished.

Furthermore, Table 6 shows time comparison between as-is and to be model as a result of substituting the conventional technology with hybrid technology. The production time decrease from 72 hours to 48 hours or equivalent with 33.33% which also in line with reduction of business process activities shown in Table 7. Devil's quadrangle measurement for cost and time in Table 8 shows the increasing cost and time variables in the proposed business process model.

Table 6. Time analysis (in hour)

Activities	As-is model	To be model
Preparing raw material and facilities	1,5	1,5
Distillation process	3	1,5
Collecting the first phase of nutmeg oil	6	2
Collecting the second phase of nutmeg oil	24	18
Collecting the third phase of nutmeg oil	36	24
Discharge	1,5	1
Total time	**72 hours**	**48 hours**

Table 7. Business process activities comparation

Activities	As-is model	To be model
Tasks	31	26
Start event	1	1
Intermediate event	2	2
End event	1	1
Time	72 hours	48 hours

Table 8. Business process measurement

Dimension	As-is model	To be model
Cost	Q ≈	Q ↑
Time	T ≈	T ↑

4. Conclusion

To conclude, the analysis of business process activities in the nutmeg oil production process reveals that:

1. Current business processes (as-is) model of nutmeg oil production process at UD. Permata Bunda consists of 31 main tasks, 1 start event, 2 intermediate events, and 1 end event.
2. Proposed business process (to be) model of nutmeg oil production process at UD. Permata Bunda consists of 26 main tasks, 1 start event, 2 intermediate events, and 1 end event.
3. Devil's quadrangle results shows that cost increase by 47.64% and time increase 33.33% as a result of substituting hybrid technology to overcome business process issue in the production process.

Reference

[1] Supriyanto, 2006, *Pemberdayaan Usaha Mikro, Kecil, dan Menengah (UMKM) Sebagai Salah Satu Upaya Penanggulangan Kemiskinan*, Jurnal Ekonomi Pendidikan, 3(1), 1-16.

[2] Darwanto, 2013, *Peningkatan Daya Saing UMKM Berbasis Inovasi dan Kreativitas (Strategi Penguatan Property Right Terhadap Inovasi dan Kreativitas*, Jurnal Bisnis dan Ekonomi, **20(20)**, 142 – 149.

[3] Data *UMKM Kabupaten Aceh Selatan*, Dinas Perindustrian dan Perdagangan Kabupaten Aceh Selatan, 2018

[4] Harrington H J, 1991, *Business Process Improvement: The Breakthrough Strategy for Total Quality*, Productivity, and Competitiveness. New York: McGraw-Hill Education

[5] Hende R Y L, Setiawan N Y, Mursityo Y T, 2018, *Perancangan Perbaikan Bisnis Proses Menggunakan Metode Business Process Improvement Pada Layanan Penerbitan Majalah*, **2(3)**, 1328–1336.

[6] Brand N, and Van der Kolk H, 1995, *Workflow Analysis and Design* [Deventer: Kluwer Bedrijfswetenschappen].

IOP Conf. Series: Materials Science and Engineering 1082 (2021) 012006 doi:10.1088/1757-899X/1082/1/012006

Comparison of BLOCPLAN and CORELAP algorithm for Material Handling Improvement at M Bakery

Raihan Dara Lufika[1], Prima Denny Sentia[1], Irfan Maulana[1]
[1]Industrial Engineering Department, Universitas Syiah Kuala, Darussalam, Banda Aceh, Indonesia

E-mail: primadennysentia@unsyiah.ac.id

Abstract. It is essential to have a suitable and effective layout design on a production floor. M Bakery is one of the bread industries that has been established since 2010. The core problem in the case study object is the far distance of the material handling movement. One solution to this problem is to improve the layout of facilities in the production area. The study will compare two different algorithms which are BLOCPLAN and CORELAP and choose the best one between these algorithms. The total movement moment of the current layout is 379,932 meters/year and OMH is IDR 1,215,712/year. The results of the BLOCPLAN based layout are the total movement moment of 263,767 meters/year, OMH of IDR 1,054,201/year, and an efficiency rate of 30.57%. While the results of the CORELAP based layout are the total movement moment of 418,710 meters/year, OMH of IDR 1,477,648/year, and an efficiency rate of -10.21%. From the comparison of the result of the two algorithms, the chosen proposed layout design is the BLOCPLAN based layout which has the highest efficiency rate of 30.57%.

1. Introduction

Facility design is one of the core aspects of a successful manufacturing process. The production cost is closely related to the facility design layout. The concept of facility planning successfully decreases the operation fee [1]. The characteristics of an ideal layout for material handling are topmost flexibility, coordination, volume utilisation, visibility, accessibility, and minimum distance [2]. There is no perfect layout. However, in some cases, the productivity rate increase significantly after the layout redesign. In one research, there is half distance left of travel route with the new layout [3]. While on other research, the change of the layout can reduce 55% of total movement and more than 7,000,000 IDR of material handling cost [4].

The facility layout issues remain as one of the growing field areas of research. There is a rise in the number of publications in the topics of facility layout optimisation [5]. This shows that there is a necessity to solve the problem regarding this issue in the real world. One research shows that the ineffective layout design can lead to profit loss due to the total material handling fee [6]. The issues are related to the increase of product demand, the shifting of market direction, the fluctuation of production fee, and many more. Therefore, the facility design should also reflect on the possibility of future expansion [7]. The other research discusses designing a layout that is considered the human factor for adjusting the Industrial Revolution 4.0 trend [8].

There are several methodologies to solve the problem about facility layout design including mathematical methods, CORELAP, Systematic Layout Plan (SLP), BLOCPLAN, Graph-Based Theory, and many more [5]. It is essential to choose which methodologies to solve the problem. The

Content from this work may be used under the terms of the Creative Commons Attribution 3.0 licence. Any further distribution of this work must maintain attribution to the author(s) and the title of the work, journal citation and DOI.
Published under licence by IOP Publishing Ltd

BLOCPLAN algorithm has successfully reduced the distance between facilities and the total material handling cost [9]. Then, CORELAP which has a similar concept aims to minimise the total displacement distance and material handling fee [10].

One of the problem in the M Bakery is the far distance despite the busy movement between the half-finished product storage and the final product storage. The other problem is the unbalance space between one aisle to another aisle between facilities. Furthermore, there are three types of product which two of them shares the same production process while the other one share a different one. This study aims to examine the current layout in the M Bakery and choosing the best layout recommendation between one using BLOCPLAN or the other one using CORELAP.

2. Methodology

2.1. Data Collection

This step is to collect the primary and secondary data for the research. The primary data is including the current layout and the workstation distance, the production floor space, the frequency of movement, the proximity level, and the production movement time. The secondary data are the production capacity, the facility number, and the employee salary fee.

2.2. Data Processing

There are fifteen processes including,

1. Current layout design
2. Calculation of distance between facilities based on the current layout
3. Calculation of the total movement moment based on the current layout
4. Calculation of total material handling cost based on the current layout
5. Activity Relationship Chart (ARC) design
6. Software BLOCPLAN 90 running
7. Block Layout design based on the Software BLOCPLAN 90 result
8. Calculation of distance between facilities based on the Software BLOCPLAN 90 result
9. Calculation of the total movement moment based on the Software BLOCPLAN 90 result
10. Calculation of total material handling cost based on the Software BLOCPLAN 90 result
11. Software CORELAP 01 running
12. Block Layout design based on the Software CORELAP 01 result
13. Calculation of distance between facilities based on the Software CORELAP 01 result
14. Calculation of the total movement moment based on the Software CORELAP 01 result
15. Calculation of total material handling cost based on the Software CORELAP 01 result

The aim of the comparison between the two results of the algorithm is to find the best result regarding the minimum distance between facilities and the minimum material handling cost. Then the layout design of the best result will be the suggestion for the company for their improvement.

3. Results and discussion

3.1. Current layout design and calculation

Figure 1 shows the current layout. The layout design refers to the observation during the production process in M Bakery's production floor. Then the design used as the reference for the calculation of the distance between facilities, total distance moment, and total material handling cost (table 1).

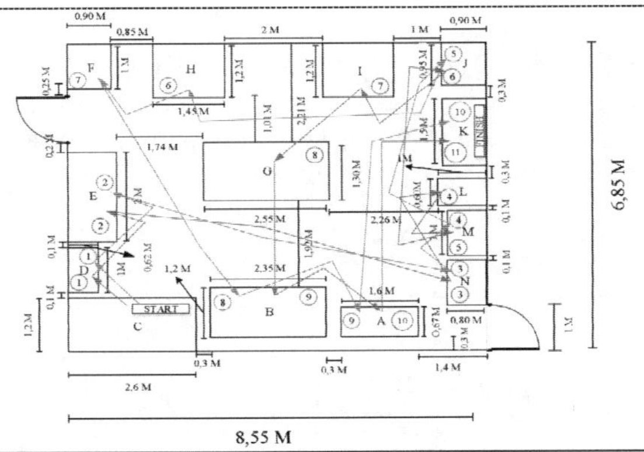

Legends:
A = Half-finished product storage
B = Packaging table
C = Raw material storage
D = Dough scale table
E = Dough mixing machine
F = Cooling table II
G = Cooling table I
H = Oven machine I
I = Oven machine II
J = Dough rising area
K = Finished product storage
L = Rolling machine
M = Dough moulding table
N = Dough cutting table

Colour:
1. Red = Production line bread A and B
2. Green = Production line bread C

Figure 1. Current layout design.

Tabel 1. Calculation of distance between facilities, total distance moment, and total material handling cost based on current layout.

No	Start Work-station	End Work-station	Worker	Movement Frequency/ Year (times)	Distance Between Facilities dij (m)	Total Movement Moment/ Year (times)	OMH/ Meter	Total OMH/ year
1	C	D	Nanda	12,045	0.21	2,529	8.55	21,622
2	D	E		12,045	1.79	21,561	8.55	184,346
3	E	N	Riki, Herman	12,045	5.65	68,054	2.94	200,078
4	N	L	Zami	9,490	2.00	18,980	3.35	63,583
5	N	M	Riki and Herman	1,825	1.20	2,190	2.94	6,438
6	L	M		6,570	0.80	5,256	2.94	15,452
7	M	J	Zami	8,395	3.72	31,229	3.35	104,617
8	J	H	Roji	2,190	5.75	12,593	5.73	72,157
9	J	I		7,300	2.30	16,790	5.73	96,206
10	I	G	Tama	10,950	4.15	45,443	3.30	149,961
11	G	B	Riki, Herman	10,950	3.06	33,507	2.94	98,510
12	H	F	Tama	2,920	1.92	5,606	3.30	18,499
13	F	B	Riki, Herman	2,920	1.84	5,373	2.94	15,796
14	B	A	Fahmi	13,870	2.03	28,156	1.52	42,797
15	A	K		13,870	5.96	82,665	1.52	125,650
TOTAL				**127,385**	**42.38**	**379,932**	**59.60**	**1,215,712**

3.2. Activity Relationship Chart (ARC) design

The aim of designing the Activity Relationship Chart (ARC) is to connect activity in pair for understanding the relationship between activities in the production floor of M Bakery. The symbols of each relationship between facilities are A, E, I, O, U, and X with the reason for close or far distance at each facility.

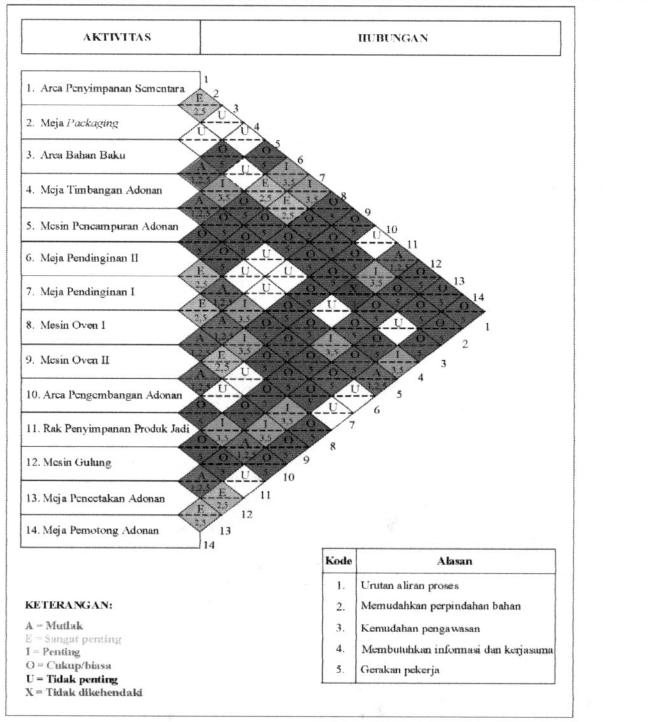

Figure 2. Activity Relation Chart(ARC).

3.3. BLOCPLAN layout design and calculation

For running the BLOCPLAN 90 software, the input data are the total floor space and Activity Relationship Chart (ARC). The result of twenty iterations for looking at the highest r-score is layout number 9 with the score of 0,76.

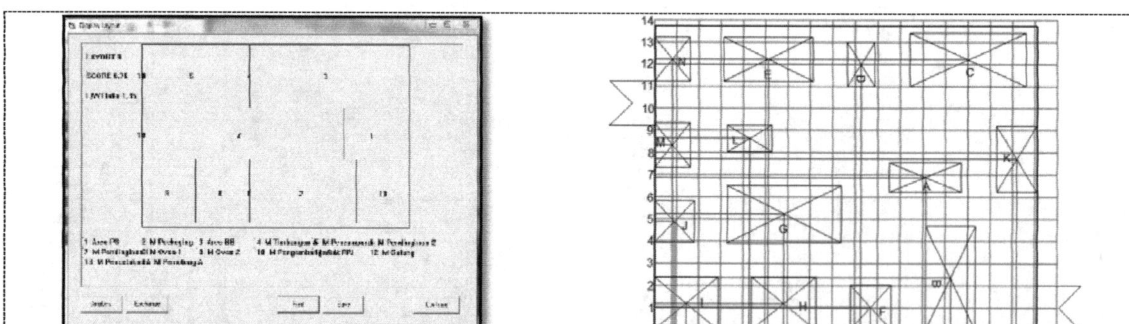

Figure 3. BLOCPLAN result (left) and block layout (right).

Tabel 2. Calculation of distance between facilities, total distance moment, and total material handling cost based on BLOCPLAN result.

No	Start Work-station	End Work-station	Worker	Movement Frequency/ Year (times)	Distance Between Facilities dij (m)	Total Movement Moment/ Year (times	OMH/Meter	Total OMH/ year
1	C	D	Nanda	12,045	2.47	29,751	8.55	254,372
2	D	E		12,045	1.97	23,729	8.55	202,880
3	E	N	Riki, Herman	12,045	2.16	26,017	2.94	76,490
4	N	L	Zami	9,490	0.02	190	3.35	636
5	N	M	Riki and Herman	1,825	1.93	3,522	2.94	10,355
6	L	M		6,570	1.91	12,549	2.94	36,893
7	M	J	Zami	8,395	1.68	14,104	3.35	47,247
8	J	H	Roji	2,190	0.75	1,642	5.73	9,411
9	J	I		7,300	1.57	11,461	5.73	65,671
10	I	G	Tama	10,950	4.21	46,100	3.30	152,128
11	G	B	Riki, Herman	10,950	2.26	24,747	2.94	72,756
12	H	F	Tama	2,920	1.73	5,051	3.30	16,670
13	F	B	Riki, Herman	2,920	2.42	7,066	2.94	20,775
14	B	A	Fahmi	13,870	0.01	139	1.52	211
15	A	K		13,870	4.16	57,699	1.52	87,703
TOTAL				**127,385**	**29.25**	**263,767**	**59.60**	**1,054,201**

Table two shows that the calculation of total material handling based on BLOCPLAN per year is 1,054,201 IDR. The farthest distance is the distance of half-finished product storage (A) to the final product storage (K) which is 57,699 meters/year. Meanwhile, the highest material handling cost per year is for the activity from the raw material area (C) to dough scaling table (D) which is IDR 254,372 OMH/year.

3.4. CORELAP layout design and calculation

The CORELAP layout designing process is including the layout drawing, coordinate point determination, the distance between facilities determination, total movement moment calculation, and total material handling cost calculation. There is only one result from the CORELAP 01 software which based on the value of Total Closeness Rating (TCR). That one result is the core data for thirteen iterations to determine which facilities should be the first one laying in the production floor.

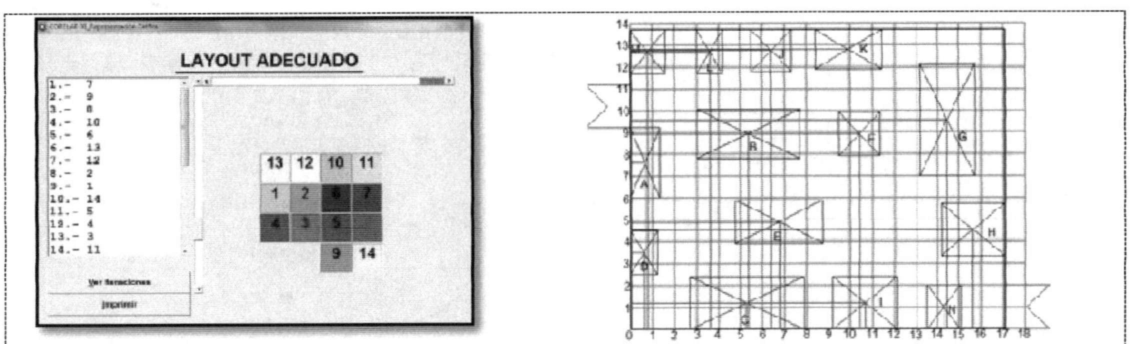

Figure 4. CORELAP result (left) and block layout (right).

Tabel 3. Calculation of distance between facilities, total distance moment, and total material handling cost based on CORELAP result.

No	Start Work-station	End Work-station	Worker	Movement Frequency/ Year (times)	Distance Between Facilities dij (m)	Total Movement Moment/ Year (times)	OMH/Meter	Total OMH/ year
1	C	D	Nanda	12,045	1.18	14,213	8.55	121,522
2	D	E		12,045	3.78	45,530	8.55	389,282
3	E	N	Riki, Herman	12,045	1.81	21,801	2.94	64,096
4	N	L	Zami	9,490	0.50	4,745	3.35	15,896
5	N	M	Riki and Herman	1,825	0.90	1,643	2.94	4,829
6	L	M		6,570	1.40	9,198	2.94	27,042
7	M	J	Zami	8,395	2.82	23,674	3.35	79,307
8	J	H	Roji	2,190	0.53	1,161	5.73	6,651
9	J	I		7,300	3.63	26,499	5.73	151,839
10	I	G	Tama	10,950	6.06	66,357	3.30	281,978
11	G	B	Riki, Herman	10,950	4.86	53,217	2.94	156,458
12	H	F	Tama	2,920	0.43	1,256	3.30	4,143
13	F	B	Riki, Herman	2,920	2.53	7,388	2.94	21,719
14	B	A	Fahmi	13,870	3.01	41,749	1.52	63,458
15	A	K		13,870	7.23	100,280	1.52	152,425
	TOTAL			**127,385**	**40.67**	**418,710**	**59.60**	**1,477,648**

Table 3 shows that the total material handling cost per year is IDR 1.477.648. The farthest distance is the distance of half-finished product storage (A) to the final product storage (K) which is 100,280 meters/year. Meanwhile, the highest material handling cost per year is for the activity from dough scaling table (D) to dough mixing machine (E) which is IDR 389,282 OMH/year.

3.5. Comparison between BLOCPLAN and CORELAP

Table 4 shows the comparison between the current layout, BLOCPLAN based layout, and CORELAP based layout. The total movement moment of BLOCPLAN algorithm and the total material handling cost is 263.767 times/year and 1.054.201 OMH/year respectively. Therefore, the efficiency rate for this algorithm is 30.57% than the current layout. Meanwhile, the total movement moment of CORELAP algorithm is 418.710 times/year and the total material handling cost is 1.477.648 OMH/year. Therefore, the efficiency rate for this algorithm is -10.21% than the current layout.

Table 4. Comparison of current layout, BLOCPLAN based layout, and CORELAP based layout.

No	Layout	Total Movement Moment/year	Total OMH/year	Efficiency Rate
1	Current layout	379,932	1,215,712	-
2	BLOCPLAN based layout	263,767	1,054,201	30,57%
3	CORELAP based layout	418,710	1,477,648	-10,21%

The main reason for the large gap between the two algorithms is because of the high rectilinear distance between the half-finished product storage to the finished product storage of CORELAP algorithm compare to others. The other reason is the result of CORELAP algorithm shows that there is an inefficiency process flow where there is a back and forth movement in the area of packaging table and cooling table two. This inefficiency may disturb the whole production process which can lead to longer production time and higher material handling cost.

4. Summary

Based on the comparison between the two algorithms, the best-proposed layout improvement is the BLOCPLAN based layout. That is because the calculation shows that the layout proposed a minimum total movement moment, a minimum total material handling fee, and the highest efficiency rate than the CORELAP based layout. It is highly possible for the M Bakery to applies the modification to the production floor because all facilities is a moveable machine, tables, and storages.

5. References

[1] Kovacs G, and Kot S, 2017, *Facility Layout Redesign for Efficiency Imrpovement and Cost Reduction*, Journal of Apllied Mathematics and Computational Mechanics, **63-74**

[2] Shanthpure M, and Shivakumar M R, 2017, *Designing Facility layout and effective material handling for a ceramic insulator plant*, International Research Journal of Engineering and Technology, **1675-1678**

[3] Nyati V, Jaybhaye M D, and Sardar V, 2017, *Optimization of Facility Layout for Imrpovement in Productivity*, Conference: 4th International Conference on Industrial Engineering (ICIE 2017)

[4] Puspita I A, Iqbal M, Pratami D, and Pratomo A, 2015, *Production Facility Layout Design Using Blocplan Algorithm*, American Scientific Publishers: Advanced Science Letters, **1-5**

[5] Kikolski M, and Ko C H, 2018, *Facility Layout Design - Review of Current Research Direction*, Engineering Management in Production and Services, **70-79**

[6] Lekan O K, Kayode O I, and Morenikeji A A, 2017, *Analysis of Plant Layout Design for Operational Efficiency with Craft Algorithms*, Acta Universitas Danubius Economica, **15-27**

[7] Kadane S M, and Bhatwadekar S, 2011, *Manufacturing Facility Layout Design and Optimization Using Simulation*, International Journal of Advanced Manufacturing System, **59-65**

[8] Li J, Tan X, And Li J, 2018, *Reserach on Dynamic Facility Layout Problem of Manufacturing Unit Considering Human Factors*, Hindawi Mathematical Problem in Engineering, **1-13**

[9] Setiawati L, Noviyarsi, and Wulandari R, 2012, *Perbaikan Tata Letak Fasilitas Produksi Dengan Menggunakan Algoritma Blocplan (Production Facility Layout Improvement using the BLOCPLAN Algorithm)*, Jurnal Teknik Industri - Universitas Bung Hatta, **206-216**

[10] Rajesh M, Naidu N V R, and Kumar P N, 2016, *Plant Layout Optimization of Oven Manufacturing Unit Using CORELAP Algorithm*, International Journal of Research in Engineering and Technology, **67-71**

IOP Conf. Series: Materials Science and Engineering 1082 (2021) 012007 doi:10.1088/1757-899X/1082/1/012007

The potential of cold energy extraction from a power plant regasification terminal for cold storage application

Sarwo Edhy Sofyan[1,9*], Wahyu Setiadi[1,9], Heru Fahlevi[2,9], Syafruddin[3,9], Purwandy Hasibuan[4,9], Taufiq Saidi[4,9], Nur Fadli[5,9], Rianjuanda[5,9], Muhammad Syukri[6,9], Teuku Mukhriza[7,9], Akram Tamlicha[1,9], and Rahmad Dawood[8,9]

[1] Department of Mechanical Engineering, Universitas Syiah Kuala, Indonesia
[2] Department of Accounting, Universitas Syiah Kuala, Indonesia
[3] Department of Management, Universitas Syiah Kuala, Indonesia
[4] Department of Civil Engineering, Universitas Syiah Kuala, Indonesia
[5] Department of Marine Science, Universitas Syiah Kuala, Indonesia
[6] Department of Physics, Universitas Syiah Kuala, Indonesia
[7] Department of Chemical Engineering, Universitas Syiah Kuala, Indonesia
[8] Department of Electrical and Computer Engineering, Universitas Syiah Kuala, Indonesia
[9] Integrated Laboratory, Universitas Syiah Kuala, Indonesia

E-mail: sarwo.edhy@unsyiah.ac.id

Abstract. Cold storage is a chamber used for food preservation. The food is stored inside the cold storage at a specific temperature and relative humidity to slow down the deterioration process caused by pathogenic organisms. The type of food stored requires specific thermal conditions e.g 0-5 °C with RH of 80-90% for vegetables, 4-5 °C for milk processing, and -25 to -45 °C for quick freezing of fish. Cold energy from the LNG regasification process is a promising option for cold storage application. Normally, prior to supply to industrial users e.g power plants, the LNG is regasified by dissipating its cold energy to another working fluid of vapourizer e.g seawater in the Open Rack Vapourizer (ORV) or ambient air in the Ambient Air Vapourizer (AAV). Recovering this cold energy to cool the cold storage can significantly reduce the energy consumption in the cooling process. Simultaneously, it reduces greenhouse gas emissions. This study aims to investigate the potential of wasted LNG cold energy in a power plant's regasification terminal, located in Aceh, as a source of energy to cool cold storage. The cooling load of cold storage under four different capacity namely 5, 10, 15, and 20 tons of fish is calculated and compared with the potential cold energy available. This study then concludes the cold storage capacity that meets the available LNG cold energy.

1. Introduction

Cold storage is a chamber used for food preservation. The food is stored inside the cold storage at a specific temperature and relative humidity in order to slow down the deterioration process caused by pathogenic organisms. The type of food stored requires specific thermal conditions e.g 0-5 °C with RH of 80-90% for vegetables, 4-5 °C for milk processing, and -25 to -45 °C for quick freezing of fish.

Content from this work may be used under the terms of the Creative Commons Attribution 3.0 licence. Any further distribution of this work must maintain attribution to the author(s) and the title of the work, journal citation and DOI.
Published under licence by IOP Publishing Ltd

Generally, there are two commonly used refrigeration cycles for cold storage, either the vapour compression refrigeration system[1] or absorption refrigeration system [2]. The former is by far the most popular and widely used in refrigeration. The system uses refrigerant, such as NH3, R-11, R-12, R-32, and R-410. The system mainly consists of four components namely compressor, condenser, expansion valve, and evaporator. The latter is a thermally driven system. The absorption system differs from vapour compression system only in the method employed for compressing the refrigerant. In which, the absorber, generator and pump are employed to replace the compressor as used at the vapour compression refrigeration system.

Cold energy derived from LNG regasification process is a promising option for cold storage application. Normally, prior to supply to industrial users e.g power plants, the LNG is regasified in the regasification facility by dissipating its cold energy either to seawater in the Open Rack Vapourizer (ORV) or to ambient air in the Ambient Air Vapourizer (AAV). Recovering this cold energy to cool the cold storage can significantly reduce the energy consumption in the cooling process. Simultaneously, it reduces greenhouse gas emissions.

There have been several studies on LNG cold energy applications. Dispenza et al. [3] presented an exergy analysis on the LNG regasification process. The performance of power cycle utilizing low grade heat input and cold energy of the LNG has been analysed by Qiang et al. [4]. Zhao et al.[5] investigated the dynamic exergy of a hybrid power system consisting of a Brayton cycle, four Rankine cycles, and a natural gas turbine expander. Ayachi et al [6] discussed the LNG cold energy for cooling applications of both data centres and vehicles. Instead of direct cooling, the LNG's cold energy is first used for air liquefaction. This liquefied air is then used as a heat transfer medium to absorb heat of data centre and vehicle compartment. From the existing literature, there is still limited study on the application of LNG cold energy for a cold storage. This study aims to investigate the potential of wasted LNG cold energy of a power plant's regasification terminal, located in Aceh, as a source of energy to cool the cold storage.

2. Cold Energy Source

At the LNG regasification facility, the LNG cold energy is dissipated either into sea water (in the ORV) or ambient air (in the AAV) increasing the temperature of LNG. Once it reaches saturation temperature, the LNG starts to boil and changes it phase from liquid into gas. The dissipated cold energy in the regasification process has not yet optimally used even though some studies report the application of LNG cold energy for cryogenic power generation cycles, oxygen liquefaction, cryogenic carbon dioxide capture [7]. This study investigates the potential of wasted LNG cold energy from regasification facility of a gas-fired power plant located in Krueng Raya, Aceh, for cold storage application.

Figure 1, shows the schematic of the receiving, storing, and regasification facility of the gas-fired power plant. As shown in the figure that the liquid LNG is unloaded from tanker trucks and circulated into a storage tank using two pump units. Subsequently, LNG is circulated from the LNG tank to the ambient air vapourizer for regasification process before feeding to the gas turbines. Table 1 summaries the operation parameters of the regasification terminal of gas-fired power plant.

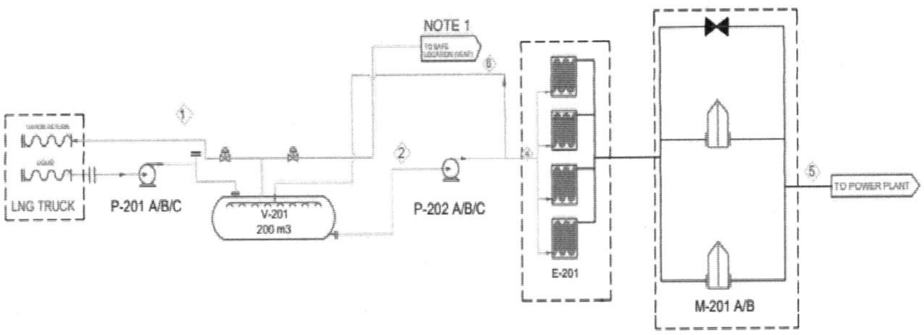

Figure 1. Schematic of regasification facility of the gas-fired power plant

Table 1. Operation parameters of the regasification terminal

Stream	1	2	3	4	5	6
Mass flow rate (kg/h)	15260	15260	2741	12519	12519	2741
Phase	Liquid	Liquid	Liquid	Liquid	Vapour	Vapour
Temperature (°C)	-162	-160	-160	-158	25	-81
Pressure (barg)	3.00	3.50	11.00	11.00	9.50	3.50
Volume flow rate (m^3/hr)	35.00	35.00	6.29	28.71	16840	3687 (STP)

Courtesy of PT. Perta Arun Gas

As can be seen from Figure 1 that the LNG enters the regasification plant (point 4) at a temperature of -158 °C and exits at point 5 as vapour at a temperature of 25 °C. During the regasification process at the ambient air vapourizer, the heat transfer takes place as the ambient air passing through the outer side of the heat exchanger. The heat transfer rate is calculated as follows:

$$Q_{reg} = \dot{m}c_g \,\Delta T + \dot{m}\, h_{fg} + \dot{m}c_f \,\Delta T \quad (1)$$

where Q_{reg} is the heat transfer rate from the regasification process (kW), $\dot{m}$ is the gas mass flow rate (kg/s), c_g and c_f are the specific heat of natural gas at gas and liquid phases (kJ/kg.K) respectively, ΔT is the temperature difference (K), and h_{fg} is the latent heat (kJ/kg).

3. Cooling Load

The cooling load of a cold storage is the summation of both total latent and sensible heat. The equations to calculate the cooling load are summarized as follows:

3.1 Total Sensible Heat (TSH)

The equation to calculate total sensible heat is summarised as:

$$TSH = RSH + OASH + RSHS \quad (2)$$

where TSH is the total sensible heat (kcal/h), RSH is the room sensible heat (kcal/h), and $OASH$ is the outside air sensible heat (kcal/h), and $RSHS$ is the room sensible heat supplement (kcal/h).

RSH is the amount of sensible load in the room plus the safety factor. This safety factor is introduced to consider the friction and leakage occur in the air circulation system. In this study, the safety factor assumes 7 % of the sensible load in the room. The sensible load in the room consists of the sensible heat of product, sensible heat gain through the wall, floor and ceiling structures, also sensible heat due to solar radiation and internal heat gain from lighting.

The sensible heat of the product is calculated as:

$$Q_{sp} = \frac{mc_p(T_1 - T_2)}{Ff} \quad (3)$$

where Q_{sp} is the product load (kcal/h), m is the mass of product (kg), c_p is the product specific heat (kcal/kg.K), T_1 and T_2 are the initial and final temperature of product (K) respectively, F is the chilling period (h), and f is the correction factor for cooling (f=1).
The sensible heat gain through the structures is presented as:

$$Q_{s\,str} = UA\Delta T \quad (4)$$

where $Q_{s\,str}$ is the heat transmits through the structure (kcal/h), U is the overall heat transfer coefficient (kcal/m^2.h), A is the heat transfer area (m^2), ΔT is the temperature difference between indoor and outdoor (K).

The sensible heat due to solar radiation is assessed as:

$$Q_{s\,r} = UA(CLTD) \tag{5}$$

where $Q_{s\,r}$ is the sensible heat due to solar radiation (kcal/h), U is the overall heat transfer coefficient (kcal/m^2.h), A is the heat transfer area (m^2), and $CLTD$ is the cooling load temperature difference (K).

The sensible heat due to internal heat gain (from lighting) is obtained from the Equation (6) as:

$$Q_{s\,I} = N_L\, P_L\, DF\, SF \tag{6}$$

where N_L is the number of lamp, P_L is the power of the lamp (W), DF is the diversity factor (DF=1.2), SF is the storage factor (3.4 Btu/Wh).

RSHS is assumed to 10 per cent of RSH, while OASH is neglected since the infiltration is considered insignificant.

3.2 Total Latent Heat (TLH)

The equation to assess the total latent heat is given as:

$$TLH = RLH + OALH + RLHS \tag{7}$$

where TLH is the total latent heat (kcal/h), RLH is the room latent heat (kcal/h), $OALH$ is the outside air latent heat (kcal/h), and $RLHS$ is the room latent heat supplement (kcal/h).

RLH corresponds to the amount of latent load in the room plus the safety factor assumed to 7 per cent of latent load in the room, while $RLHS$ is equal to 10 per cent of RLH and $OALH$ is neglected.

The latent load product due to the respiration process is given as:

$$Q_{res} = \frac{mQ}{F} \tag{8}$$

where Q_{res} is the latent load due to product respiration (kcal/h), m is the mass of the product (kg), Q is the heat removed from the product (kcal/kg), and F is the cooling period (h).

ASHRAE [8] specifies the fish's freezing temperature is -2.2 °C, the specific heats of fish above and below freezing point are 3.18 kJ/kg.K, and 1.71 kJ/kg.K, respectively, and the latent heat of fish is 233.9 kJ/kg.

3.3 Grand Total Heat (GTH)

The grand total heat is the sum of both the total sensible and latent heat as presented by Equation (9). Tables 2-3 present the design parameters and wall material properties of the cold storage.

$$GTH = TSH + TLH \tag{9}$$

Tabel 2. Design parameter of the cold storage

Design Parameter	Value
Ambient air temperature (°C)	27.5
Outdoor air relative humidity (%)	75
Wind speed (mph)	7.5
Indoor air temperature (°C)	-40
Indoor air relative humidity (%)	90

Tabel 3. Wall thermal conductivity of the cold storage

Material	Thickness (m)	Thermal conductivity (Kkal/h °C)	Material	Thickness (m)	Thermal conductivity (Kkal/h °C)
Wall			**Ceiling**		
Carbon steel	0.002	30.95	Carbon steel	0.002	30.95
Glass wool	0.03	0.032	Glass wool	0.045	0.032
Polystyrene	0.05	0.033	Polystyrene	0.1	0.033
Air gap	0.11	0.095	Air gap	0.21	0.095
Aluminum	0.002	176	Aluminum	0.02	176
Floor			**Door**		
Asphalt shingles	0.005	0.053	Carbon steel	0.002	30.95
Pine wood board	0.005	0.126	Glass wool	0.03	0.032
Polystyrene	0.1	0.033	Polystyrene	0.031	0.033
Glass wool	0.04	0.032	Air gap	0.03	0.095
Carbon steel	0.004	30.95	Aluminum	0.002	176

4. Result and Discussion

Table 4 presents the potential of cold energy that could be harvested during the regasification process based on Equation 1. As can be seen that the potential of wasted energy that can be extracted is enormous. Increasing the natural gas temperature to -60 °C at a constant flow rate of 12519 kg/h can generate 2151.6 kW of cold energy. This substantial amount of energy is derived from both sensible and latent heat during the regasification process. From the pressure-enthalpy diagram, it was found that the latent heat of vapouration of the LNG is 410 kJ/kg at the pressure of 11 bar.

Table 4. Cooling capacity of during the regasification process

Flow Rate (kg/h)	Specific Heat (kJ/kgK)	Inlet Temperature (°C)	Outlet Temperature (°C)	Latent Heat (kJ/kg)	Cooling Capacity (kW)
12519	2.13	-158	-120	-410	-1707.2
12519	2.13	-158	-110	-410	-1781.3
12519	2.13	-158	-100	-410	-1855.4
12519	2.13	-158	-90	-410	-1929.4
12519	2.13	-158	-80	-410	-2003.5
12519	2.13	-158	-70	-410	-2077.6
12519	2.13	-158	-60	-410	-2151.6

The cooling load for four different capacity of cold storage namely 5, 10, 15, and 20 tons of fish is calculated and presented in Table 5. As can be observed from the table that the cooling load of cold storage for 5 tons capacity of fish is 177.3 kW. The value of the cooling load increases as the capacity of the cold storage increases. At the capacity of 20 tons, the cooling load of the cold storage is 705.5 kW, which is still far below the available cooling capacity. Varying the mass flow rate of the LNG enables us to control the cooling capacity that satisfies the cooling load. Future work includes designing larger cold storage to make use of the substantial of cold energy that is still extractable.

Table 5. Cooling load of four different capacity of cold storage

Mass of Product (kg)	Temperature		Dimension			Cooling Period (h)	TSH (kW)	TLH (kW)	GTH (kW)
	T_i Product (°C)	T_f Product (°C)	Length (m)	Width (m)	Height (m)				
5000	25	-40	5	3	3	6	64.4	112.9	177.3
10000	25	-40	5	4	3	6	127.2	225.9	353,1
15000	25	-40	6	5	3	6	190.2	338.9	529,1
20000	25	-40	8	6	3	6	253.6	451.9	705,5

5. Conclusion

In conclusion, a study on the potential of LNG cold energy from the regasification facility of the gas-fired power plant, located in Aceh, has been presented. The result reveals that increasing the LNG temperature from -158 °C to -60 °C, at a mass flow rate of 12519 kg/h, offers the cooling capacity of 1707.2 kW. This cooling capacity is far beyond the cooling load of the cold storage with a capacity of 20 tons of fish. Future work includes sizing the cold storage to extract the maximum potential of cold energy available.

References

[1] Badr O, O'Callaghan P W, and Probert S D, 1990, *Vapour-compression refrigeration systems*, Appl. Energy, **36**, 303–331, doi: 10.1016/0306-2619(90)90004-W.

[2] Sofyan S E, Farhan M, Khairil, Jalaluddin, and Akram, 2020, Theoretical Study of the Absorption Refrigeration Cycle Using Water-Lithium Bromide as Working Pair for Cold Storage Application, IOP Conf. Ser. Mater. Sci. Eng., **796(1)**, doi: 10.1088/1757-899X/796/1/012015.

[3] Dispenza C, Dispenza G, La Rocca V, and Panno G, 2009, *Exergy recovery in regasification facilities - Cold utilization: A modular unit*, Appl. Therm. Eng., **29(1)**, 3595–3608, doi: 10.1016/j.applthermaleng.2009.06.016.

[4] Qiang W, Yanzhong L, and Jiang W, 2004, *Analysis of power cycle based on cold energy of liquefied natural gas and low-grade heat source*, Appl. Therm. Eng., **24(4)** 539–548, doi: 10.1016/j.applthermaleng.2003.09.010.

[5] Zhao L, Zhang J, Wang X, Feng J, Dong H, and Kong X, 2020, *Dynamic exergy analysis of a novel LNG cold energy utilization system combined with cold, heat and power*, Energy, **212**, 118649, doi: 10.1016/j.energy.2020.118649.

[6] Ayachi F, Lizhong Y, Magro F D, Meneghetti A, and Romagnoli A, 2019, *Assessment of LNG cold energy utilization for road vehicles and data-centres cooling using liquid air*, Energy Procedia, **158**, 5047–5052, doi: 10.1016/j.egypro.2019.01.656.

[7] He T, Chong Z R, Zheng J, Ju Y, and Linga P, 2019, *LNG cold energy utilization: Prospects and challenges*, Energy, **170**, 557–568, doi: 10.1016/j.energy.2018.12.170.

[8] ASHRAE, 2018 Ashrae Handbook.

Working posture analysis of wall building activities in construction works using the OWAS method

M Iqbal[1]*, **L Angriani[1]**, **I Hasanuddin[1]**, **F Erwan[1]**, **H Soewardi[2]** and **A Hassan[3]**

[1]Department of Mechanical and Industrial Engineering, Universitas Syiah Kuala, Banda Aceh, Indonesia
[2]Department of Industrial Engineering, Indonesian Islamic University, Yogyakarta, Indonesia
[3]Electronic & Automation Section, Universiti Kuala Lumpur, Malaysian Spanish Institute, Kulim, Kedah, Malaysia

*E-mail: mohd.iqbal@unsyiah.ac.id

Abstract. Activities of construction works are carried out manually and continuously with critical and repetitive working postures. The poor working condition causes fatigue, musculoskeletal disorder, in the worst case could lead to working accident. The paper discussed critical working postures of construction works of house construction project in Indonesia. Three mains activities have been identified to be studied, namely building up the brick wall, plastering the wall, and casting the concrete column. The activities were video recorded, the dimension of the work place and its component were measured and further analyzed in ergonomics laboratory. The postures were studied using Ovako Working Posture Analysis System (OWAS) method to identify their critical levels. The improvements were proposed to remove the critical posture and reduce the risk of injury. Result of the study showed that building up the brick wall was the most critical task since it has 2 critical postures, while plastering the wall and casting the concrete column have 1 critical posture respectively. The propose improvement has successfully removed the critical posture and reduce the risk of injury.

1. Introduction

Construction works consist of many groups of activities. According to parts of the building, generally the activities starts from foundation works, wall building and roof installation. The foundation is the most important part of the building. As the base of the building, the foundation must be able to receive the load of the hole construction. With regard to this function, the foundation is the strongest part of the building. The wall protects the building from outside disturbance, such as wind, rain, snow, sun light and hot or cold temperature. The wall also plays important role for the access from outside to the inside of the building or the other way rounds. The windows and doors are installed in the wall for access purpose. The wall contributes to the esthetic of the building most among all component of the building. That is why the construction of the wall takes serious attention by the contractor, workers, consultant and the owner of the building. The roof is the top cover of the building. Similar to the wall, the roof protects the building from the outside disturbance, especially that come from the above of the

Content from this work may be used under the terms of the Creative Commons Attribution 3.0 licence. Any further distribution of this work must maintain attribution to the author(s) and the title of the work, journal citation and DOI.
Published under licence by IOP Publishing Ltd

building, such as wind, rain, snow and sun light. The roof must be strong enough to resist from the load of outside disturbance. The roof also contributes to the esthetic of the building. That is why the design of the roof must meet the strength and appearance requirement [1, 2].

Managing the construction works is a unique task. Most of the work activities are depend one each other. Many activities could be started only after the other activities are completed. Dealing with this problem, the project manager usually manages the works according to the part of the building, starts from foundation, continue to the wall and finish with the roof. However, in some works, the activities take a shorter time compare to the completing non-activity process. For example, in the concrete casting process, after only 3 hours working to prepare the cement and cast it into the mold, the workers have to wait up to 12 hours for natural curing process of the concrete. In this case, in order to maintain the efficiency, the activities are managed according the nature of the task, such as concrete work, iron work, wood work, roof installation and painting. The workers could continue to the other part of the building while the non-activity natural finish process to be completed [3, 4].

The unique management described above make the construction works a difficult task to the workers. They have to move from one location to the other location of the construction area during the work. They have to return to the previous location frequently, to check whether the non-activity process have been completed and the following tasks could be started. Walking in a construction area put the worker in a risk of accident since many harmful materials are presented. The construction works are considered as heavy works as well. The workers need to carry heavy tools, equipment and material, as well as to control/ drive heavy machineries. Most of the activities are provided outside in the open space, where the workers are exposed to the hot temperature (from the machines and sun light), cold temperature (at night or snow) and extreme weather (wind or rain). All of this working condition requires a serious safety consideration for the construction workers. The working environment as well as the working task need to be improved in order to increase the safety and ergonomics level for the construction workers [5, 6].

Many researches have been conducted to study and improve the safety and ergonomics level of construction work. Lee and Han [7] studied the critical working posture provided by the worker when tying beams with steel bars, assembling column templates, and cement grouting. They identified that bent and twisted trunk posture was the major poor posture for construction workers. Construction workers were required to handle heavy load as reported by Lunde et al [8]. Handling a heavy load during a long duration of industrial working led to muscle injury and decreases the working productivity and injury [9]. Lunde et al [8] evaluated the level of cardiovascular load in relation to individual factors, work ability, musculoskeletal pain and subjective general health. They concluded that the cardiovascular load at work was significantly associated with age and oxygen consumption, but not with workability, musculoskeletal pain or subjective general health.

Handling heavy load during the work must be minimized, either by minimizing the load weight, shorten the handling time or reducing the handling frequency. The individual load weight could be reduced by increasing the number of the workers [10]. The load that previously handled by one worker was shared among several workers, made the load lighter for each worker. The load could be minimized using supporting equipment such as developed by Salam et al [11]. They proposed a new design and prototype of artificial skeleton that fit user's anthropometry. The mechanism of the skeleton reduces the bio-mechanic load up to 20% of the handling load.

Working environment has been studied by researcher since it effects the worker health and working performances [12]. The study by Iskandar et al [13] found that a poor working environment, i.e. heavy dust, poor lighting and high temperature, significantly decrease the workers motivation, productivity and product quality. Social environment has been considered as significant factor that effect the worker's performances [14]. Erwan et al [15, 16] studied the burnout syndrome among the workers caused by long shift time, poor working environment and unmanageable extra works. They proposed a better working management in order to decrease the stress among the workers.

Productivity and product quality were the main concern of modern industry [17, 18]. Related production parameters need to be controlled all the time to maintain production efficiency and to reduce production cost [19, 20]. It was proven that safety and ergonomic factors are among important parameters that significantly affect workers performance that contribute directly to productivity, product quality and production cost [21]. Ergonomics research developed many new methods or adopted successful methods from other research area to increase the safety and ergonomic level in working environment. Iqbal et al [22] used factorial design and response surface methodology to optimize the design parameters of workplace in order to increase the working performance. As the result of the study, they found the optimum value of design parameters that decreased working time, decreased critical posture, increased the lifting weight limit and decreased metabolic energy of the workers while providing the work tasks. The study has been further developed using virtual engineering for easier workplace design process and better experimental result [23].

This paper reports the research on critical working posture provided by the workers in construction project. Even though such a study has been explored by many researchers around the world, construction work is a unique working activity. Unsimilar to manufacturing works, that mostly have been standard and conducted uniformly in the factories, many construction works were not yet standard. The same job, could be provided by many different ways and very much depends on the available equipment, tools, technology and cost as well as affected by gender, social level and culture [24]. This research study the wall building activities in a common house building in Indonesia. Three main activities were considered, namely building up the brick wall, plastering the wall, and casting the concrete column. All of the activities were provided manually. The critical posture of the workers was investigated and analyzed using Ovako Working Posture Analysis System (OWAS) method. The recommendation was provided to improve the workplace and/ or the work task to eliminate or reduce the critical working posture.

Table 1. OWAS score for body part and load

Body part	OWAS score	Description
Back	1	Straight
	2	Bent
	3	Twisted or bent to one side
	4	Bent and twisted or bent and bent to one side
Arms	1	Both arms below shoulder level
	2	One arm at or above shoulder level
	3	Both arm at or above shoulder level
Legs	1	Sitting, legs below seat level
	2	Standing, legs straight
	3	Standing on one leg, legs straight
	4	Standing on both leg, legs bent
	5	Standing on one leg, legs bent
	6	Kneeling on one or both knees
	7	Walking or moving
Load	1	Less or equal to 10 Kg
	2	Greater than 10 Kg and less or equal to 20 Kg
	3	Greater than 20 Kg

2. Method

The working activities were observed directly in the construction area and recorded by video tape. The dimension of the work place's component were measured and photo recorded. The video and photograph were brough to the ergonomics laboratory for further investigation. The working postures were analysed by using OWAS method. OWAS is a working posture analysis technique which was developed by a steel company named Ovako [25]. OWAS gives a score for the position of three main body parts (i.e. back, arm and leg) and load that handled by the subject, as shown in Table 1.

The score depends on the level of critical, difficulties, risk of injury and weight of the load. Higher score represents the higher level of critical posture. Accordingly, each of individual posture during the task has 4 digit OWAS score. Based on the combination of the 4 digit OWAS score, the class of posture category is given, as shown in Table 2. The class represent the critical level as posture, start from class 1 (normal) until class 4 (the most critical posture). Because of its simplicity and usefulness, OWAS method has been used in many cases for decades, i.e. perchery system [26], automotive [27], medical
[28] and musculoskeletal problem [29].

3. Result and discussion

The construction process of building a brick-cement wall requires three main activities, which is casting, arranging bricks and cement, and plastering. Activities carried out by construction worker involves critical working postures, such as scooping cement, laying bricks, sweeping cement on bricks, plastering and casting. The critical postures lead to the risk of injury of the musculoskeletal system, or in the worst-case lead to dangerous working position and probably continue to working accident.

Tabel 2. OWAS class for body posture and load

Back	Arms	Legs																				
		1			2			3			4			5			6			7		
		Load			Load			Load			Load			Load			Load			Load		
		1	2	3	1	2	3	1	2	3	1	2	3	1	2	3	1	2	3	1	2	3
1	1	1	1	1	1	1	1	1	1	1	2	2	2	2	2	2	1	1	1	1	1	1
	2	1	1	1	1	1	1	1	1	1	2	2	2	2	2	2	1	1	1	1	1	1
	3	1	1	1	1	1	1	1	1	1	2	2	3	2	2	3	1	1	1	1	1	2
2	1	2	2	3	2	2	3	2	2	3	3	3	3	3	3	3	2	2	2	2	3	3
	2	2	2	3	2	2	3	2	3	3	3	4	4	3	4	4	3	3	4	2	3	4
	3	3	3	4	2	2	3	3	3	3	3	4	4	4	4	4	4	4	4	2	3	4
3	1	1	1	1	1	1	1	1	1	2	3	3	3	4	4	4	1	1	1	1	1	1
	2	2	2	3	1	1	1	1	1	2	4	4	4	4	4	4	3	3	3	1	1	1
	3	2	2	3	1	1	1	2	3	3	4	4	4	4	4	4	4	4	4	1	1	1
4	1	2	3	3	2	2	3	2	2	3	4	4	4	4	4	4	4	4	4	2	3	4
	2	3	3	4	2	3	4	3	3	4	4	4	4	4	4	4	4	4	4	2	3	4
	3	4	4	4	2	3	4	3	3	4	4	4	4	4	4	4	4	4	4	2	3	4

INTERNASIONAL OF THE RESULT
1- No action required
2- Corrective actions required in the near future
3- Corrective actions should be done as soon as possible
4- Corrective actions for improvement required immediately

(a) (b)

(c) (d)

Figure 1. Building the brick wall, (a) scoop the cement from the contsiner, (b) apply the cement on the wall, (c) reach the brick from the floor level and (d) position the bricks on the wall

3.1. Building up the brick wall

The brick-cement wall is one of the important parts of a building. Arranging brick-cement was conducted manually by the worker starts with laying the bricks horizontally with approximately 5 cm gap in between. The gap was then filled with the cement mix. Five cm cement mix was applied on the top of the brick row, and the task continue with next row until the wall was complete. Based on the early investigation, four critical posture could be identified during the building of brick wall, i.e when the worker scooped the cement from the container, applied cement on the top surface of the wall, took the brick, and put the bricks on the wall.

When the worker scooped cement from the container located on the floor level at his right side, as shown in figure 1(a), he needed to bent and twist his back. Referring to Table 1, the OWAS score for this back position was 4. However, both of the arms were under the shoulder level, which was considered as normal position (OWAS score for arms = 1). Since the location of the cement container was too low for the worker, he needed to bend his legs, made the OWAS score for legs = 4. The weight of the cement was less then 10 kg, made the OWAS score for the load = 1. The total OWAS score for the said posture was 4-1-4-1. According to Tabel 2, it felt in class 4, which was considered as extremely critical posture and corrective actions for improvement was required immediately.

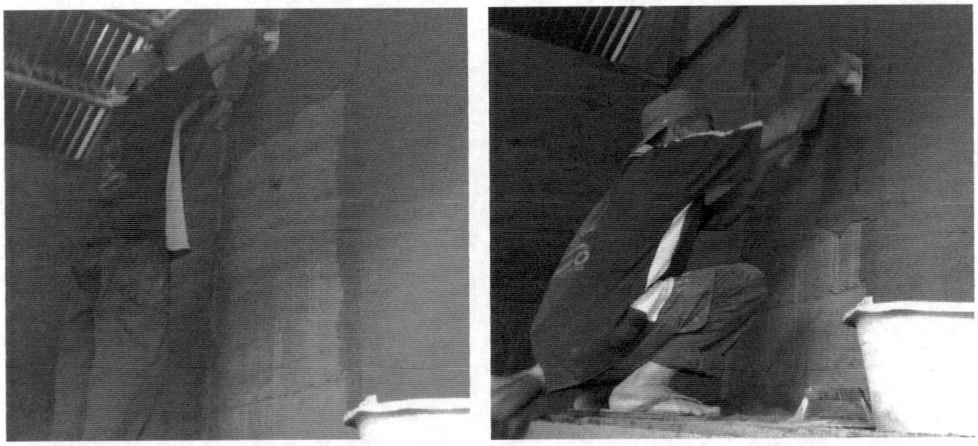

(a) (b)

Figure 2. Plastering process of the wall, (a) plastering the higher part, (b) plastering the lower part

In order to remove the critical posture, a table was provided in the right side of the worker with the high approximately similar to the worker's elbow level. The cement container was put on the table. Now the worker did not need to bent, neither to twits his back as well as to bent his legs when scooping the cement from the container. The total OWAS score for new posture was 1-1-2-1, which felt in class 1 and considered as normal and safe position.

When the worker applied cement on the top surface of the wall, as shown in Figure 1 (b), he needed to bend his back because the surface was lower than his elbow high and according to the task requirement, he needed to put his lower arm in horizontal position. The OWAS back score for this position was 2. The total OWAS score for the position was 2-1-2-1 (both of the arms were below shoulder level, standing position with legs straight, and the weight of the load was less than 10 Kg), made the position felt in class 2, which was considered as a safe position but the corrective action was required in the near future. According to Table 2, the posture could be improved if the OWAS back score changes from 2 to 1. However, the current standing position could not be changed to sitting position because of the nature of the task. As the conclusion, the working posture reminded in class 2.

When the worker took the brick from the floor level (Figure 1.c), he provided the same position as he scooped the cement (Figure 1.a). Accordingly, the similar analysis and the improvement were conducted to reduce the critical level of the working posture from class 4 (extremely critical posture) to class 1 (normal posture). Similarly, when the worker put the bricks on the wall (Figure 1.d), he provided the same position as he applied cement on the top surface of the wall (Figure 1.b). Accordingly, the similar analysis was conducted, and no improvement could be provided due to the nature of the working task and the working posture reminded in class 2.

3.2. Plastering the wall

Plastering is the process to cover the brick wall surface with cement. It was conducting manually by the worker. The cement mix was put in a container. The worker took the cement mix from the container and applied it to the surface of the wall. The task was provided during standing and kneeling depend on the location of the plastering. Changing the position from standing to kneeling frequently gave a stress to the worker's body. The working posture analysis was required to reduce the stress and remove the critical posture.

When the worker plastered the higher part of the wall (as shown in figure 2.a), his back was in normal straight position (Back = 1), but both of his arms are above his shoulder level (Arms = 3). He works in standing postion with both legs straight (Legs = 2). The weight of the load (cement mix and scoop) was

(a) (b)

(c)

Figure 3. Casting process, (a) taking the bucket of material, (b) lifting the bucket, (c) pouring the material into the column mould

less thatn 10 Kg (Load =1). The total OWAS score for the posture was 1-3-2-1, making a high wall plaster position in category / class = 1 which is a normal category and no imporvement was required.

When the worker plastered the lower part of the wall (as shown in Figure 2.b), his back was bent (Back = 2), and both arms above shoulder level (Arm = 3). He provided the task with kneeling position (Legs = 6). The weight of the load was less than 10 Kg (Load = 1). The total OWAS score obtained was 1-3-6-1, making the lower wall plaster posture in category/class = 4 which was an extremly critical posture and corrective actions for improvement required immediately. Considering the situation, a sit support could be provided to change the working position from kneeling to sitting (Legs = 1). The total OWAS score became 2-3-1-1, which felt in class 2 that much safer for the worker compare to the previous posture.

3.3. Casting of concrete column

The casting activity studied here was conducted in making the vertical concrete column. The casting process was carried out manually by the worker, starts with taking the bucket containing concrete mix (cement mixed with sand, gravel and water) as shown in Figure 3.a, continue with lifting the bucket (Figure 3.b), and finally pouring the concrete mix into the column mould (Figure 3.c). The process continued until the mould was completely full of the concrete mix. However, based on the analysis, only one critical posture were identified during the casting cycle task.

The critical posture was occurred when the worker took the bucket from the flatform, as shown in Figure 3(a). The worker needed to bent and twist his back to reach to the bucket located on the floor level at his right side. The OWAS back score for this position was 4 (refer to Table 1). However, both of the hands were under shoulder level, which was considered as normal/ safe position (Arms score = 1). Since the bucket was too low, the worker needed to bent his leg (Legs score = 4) when he reached the bucket on the floor level. The weight of the bucket with material is lower than 10 Kg, made the Load score =1. The total OWAS score was 4-1-4-1. Referring to Table 2, the working position was considered as class 4 (a dangerous critical posture and corrective actions for improvement was required immediately).

Based on the observation, the critical posture occurred because of the location of the concrete mix bucket that too low for the worker. A table could be provided in the wright side of the worker and the bucket was put on the table, made it in the same level of the worker's hand high while standing. The new position made the worker worked with his back straight and his legs straight. The new total OWAS score was 1-1-2-1, made the position felt in class 1 (normal posture).

4. Conclusion

The ergonomics assessment for construction works have been conducted. Four critical postures were found during the investigation. According to OWAS method, they were categorized as extremely critical posture and the improvement need to be provided immediately. The critical tasks have been conducted by the workers for a long time and they have been a routine daily work activity. The propose improvement has successfully remove the critical postures and reduce the risk of injury. The fact that the improvement was a simple action and did not need any significant additional cost, time and manpower, showed that the ergonomics and safety awareness in the construction project was very low. The socialization and training need to be conducted regularly for the workers and the contractor to update their understanding on Occupational Safety and Health in construction work. The law enforcement also needs to be implemented as well as direct inspection to make sure that the ergonomics and safety were seriously considered. For future work, the anthropometry of the worker will be considered to study whether the equipment used during the works match the body size of the worker. It was an important ergonomic factor since the anthropometry of the local workers was unique [30] and many equipment was imported from other countries.

Reference

[1]. Allen E, and Iano J, 2019, *Fundamentals of building construction: materials and methods*, John Wiley & Sons

[2]. Fewings P, and Henjewele C, 2019, *Construction project management: an integrated approach*, Routledge

[3]. Jaskowski P, and Biruk S, 2020, *Scheduling of repetitive construction processes with concurrent work of similarly specialized crews*, Journal of Civil Engineering and Management, **26(6)**, 579-589, https://doi.org/10.3846/jcem.2020.12914

[4]. Gurmu A T, Aibinu A A, and Toong Khuan C. A, 2016, *Study of best management practices for enhancing productivity in building projects: construction methods perspectives*, Construction Economics and building. **16(3)**, 1-19, https://doi.org/10.5130/AJCEB.v16i3.4882

[5]. Carpio-de Los Pinos A J, and González-García MD, 2020, *Development of the protocol of the occupational risk assessment method for construction works: level of preventive action*,

International journal of environmental research and public health, **17(17)**, 6369, https://doi.org/10.3390/ijerph17176369

[6]. Ding GK, 2008, *Sustainable construction-The role of environmental assessment tools*, Journal of environmental management, **86(3)**, 451-64, https://doi.org/10.1016/j.jenvman.2006.12.025

[7]. Lee T H, and Han C S, 2013, *Analysis of working postures at a construction site using the OWAS method*, International Journal of Occupational Safety and Ergonomics, **19(2)**, 245-50, https://doi.org/10.1080/10803548.2013.11076983

[8]. Lunde L K, Koch M, Veiersted K B, Moen G H, Wærsted M, and Knardahl S, 2016, *Heavy Physical Work: Cardiovascular Load in Male Construction Workers*, International Journal of Environmental Research and Public Health, **13(4)**, 356, https://doi.org/10.3390/ijerph13040356

[9]. Zeki M, Iskandar I, and Iqbal M, 2019, *Analisis Efektifitas Kerja Pengangkatan Beban Pada Bagian Pengantongan Di PT. Pupuk Krueng Geukuh*, Industrial Engineering Journal, **8(2)**, https://journal.unimal.ac.id/miej/article/view/404/299

[10]. De Looze M P, Bosch T, Krause F, Stadler KS, and O'Sullivan L W, 2016, *Exoskeletons for industrial application and their potential effects on physical work load*, Ergonomics, **59(5)**, 671-81, https://doi.org/10.1080/00140139.2015.1081988

[11]. Salam R, Iqbal M, and Hasanuddin I, 2018, *Desain dan Analisis Artificial Exoskeleton pada Prajurit TNI*, Jurnal Optimasi Sistem Industri, **17(2)**,135-42, https://doi.org/10.25077/josi.v17.n2.p135- 142.2018

[12]. Sawacha E, Naoum S, and Fong D, 1999, *Factors affecting safety performance on construction sites*, International journal of project management. **17(5)**, 309-15, https://doi.org/10.1016/S0263-7863(98)00042-8

[13]. Hasanuddin I, Iqbal M, Ghazilla R A, and Erwan F, 2019, *The Effect of Working Environment and Motivation towards Labors' Performance: A Case Study in Aceh at PT Pupuk Iskandar Muda*, Aceh International Journal of Science and Technology. **8(1)**, 35-40, https://core.ac.uk/download/pdf/290139434.pdf

[14]. Lloyd C, King R, and Chenoweth L, 2002, *Social work, stress and burnout: A review*, Journal of mental health, **11(3)**, 255-265, https://doi.org/10.1080/09638230020023642

[15]. Erwan F, Iqbal M, and Hasanuddin I, 2018, *The analysis of nursing's work systems in relation to burnout syndrome (A case study: nurses in RSUDZA, BLUD RSIA, and RSUD Meuraxa, Banda Aceh, Indonesia)*, IOP Conference Series: Materials Science and Engineering, **453(1)**, 012051, https://iopscience.iop.org/article/10.1088/1757-899X/453/1/012051/pdf

[16]. Erwan F, Iqbal M, Hasanuddin I, Zuhri S, and Maydini C M, 2020, *Burnout Syndrome analysis among hospital nurses using Maslach Burnout Inventory-Human Service Survey (MBI-HSS): a case study*, IOP Conference Series: Materials Science and Engineering, **931(1)**, 012025, https://iopscience.iop.org/article/10.1088/1757-899X/931/1/012025/pdf

[17]. Iqbal M, Bamatraf O A A. and Tadjuddin M, 2018, *The study of production performance of water heater manufacturing by using simulation method*, Journal of Physics: Conference Series, **962(1)**, 012019). https://iopscience.iop.org/article/10.1088/1742-6596/962/1/012019/pdf

[18]. Iqbal M, Purnomo, M R A, Imra M A B M, Konneh M, and Karim A N, 2012, *The study on production performances of part transporter in flexible manufacturing system by using simulation method*, Advanced Materials Research, **576**, 714-717, https://doi.org/10.4028/www.scientific.net/AMR.576.714

[19]. Karim A N, Hamdan M A H, Alam M Z, and Iqbal M, 2012, *Economic Scenario of Cellulase Enzyme Production Using Palm Oil Mill Effluent (POME)*, Advanced Materials Research, **576**, 723-726, https://doi.org/10.4028/www.scientific.net/AMR.576.723

[20]. Purnomo M R, Iqbal M, and Sufa M F, 2012, *Solving Multiple Routes Travelling Salesman Problem Using Modified Genetic Algorithm*, Advanced Materials Research, **576**, 718-722, https://doi.org/10.4028/www.scientific.net/AMR.576.718

[21]. Battini D, Faccio M, Persona A, and Sgarbossa F, 2011, *New methodological framework to improve productivity and ergonomics in assembly system design*, International Journal of industrial ergonomics, **41(1)**, 30-42, https://doi.org/10.1016/j.ergon.2010.12.001

[22]. Iqbal M, Hasanuddin I, Hassan A, Soufi M S, and Erwan F, 2020, *The Study on Ergonomic Performances Based on Workstation Design Parameters using Virtual Manufacturing Tool*, International Journal of Integrated Engineering, **12(5)**, 124-9, https://publisher.uthm.edu.my/ojs/index.php/ijie/article/view/6092

[23]. Iqbal M, Soewardi H, Hassan A, Haron C H, 2004, *Ergonomic Study for Optimum Printing Workstation Using Factorial Experiment and Response Surface Methodology*, Journal of Occupational Safety and Health, **1(2)**, 42-49

[24]. Wang D, Dai F, and Ning X, 2015, *Risk assessment of work-related musculoskeletal disorders in construction: State-of-the-art review*, Journal of Construction Engineering and management, **141(6)**, 04015008, https://ascelibrary.org/doi/abs/10.1061/(ASCE)CO.1943-7862.0000979

[25]. Karhu O, Kansi P, and Kuorinka I, 1977, *Correcting working postures in industry: A practical method for analysis*, Applied Ergonomics, **8(4)**, 199–201, https://doi.org/10.1016/0003-6870(77)90164-8

[26]. Scot G B, and Lambe N R, 2006, *Working practice in phercery system, using the OVAKO working posture analysing system (OWAS)*, Applied Ergonomics, **27(4)**, 281-284

[27]. Ismail A R, Yeo M L, Haniff M H M, Zulkifli R, Deros B M, and Makhtar N K, 2009, *Assessment of postural loading among the assembly operators: A case study at Malaysian automotive industry*, European Journal of Scientific Research, **30(2)**, 224-235

[28]. Verjans M, Schütt A, Schleer P, Struck D, and Radermacher K, 2018, *Postural workloads on paramedics during patient transport.* Current Directions in Biomedical Engineering, **4(1)**, 161-164, https://doi.org/10.1515/cdbme-2018-0040

[29]. Hellig T, Mertens A, and Brandl C, 2018, *The interaction effect of working postures on muscle activity and subjective discomfort during static working postures and its correlation with OWAS*, International Journal of Industrial Ergonomics, **68**, 25-33, https://doi.org/10.1016/j.ergon.2018.06.006

[30]. Iqbal M, Salam R, Hasanuddin I, and Hassan A B, 2020, *Body Segment Dimensions of Indonesian Male Army*, IOP Conference Series: Materials Science and Engineering, **931(1)**, 012013, https://iopscience.iop.org/article/10.1088/1757-899X/931/1/012013/meta

IOP Conf. Series: Materials Science and Engineering 1082 (2021) 012009 doi:10.1088/1757-899X/1082/1/012009

Effect of number of permanent magnetic poles on 3D printed coreless generator rotor

Safarudin Gazali Herawan[1], Said Abi Syahputra[1], Ernie Mat Tokit[2], Fatimah Al-Zahra Mohd Sa'at[2], Mohamad Afzanizam Mohd Rosli[2]

[1]Industrial Engineering Department, Faculty of Engineering, Bina Nusantara University, Jakarta, Indonesia 11480
[2]Centre for Advanced Research on Energy (CARe), Fakulti Kejuruteraan Mekanikal, Universiti Teknikal Malaysia Melaka, Hang Tuah Jaya, 76100 Durian Tunggal, Melaka, Malaysia.

Email: safarudin.gazali@binus.edu

Abstract. 3D-printed coreless generator using rotors with 4 poles permanent magnet and 8 poles permanent magnet as source of external magnetic field has been presented for energy harvesting applications. The prototype of proposed coreless generator is built using 3D-printer with Black Polylactic Acid (PLA) as a filament for stator housing, rotor and other parts. The major parts of the proposed coreless generator are the stator, which has four coils of winding copper wire and two rotors with four permanent magnet and eight permanent magnets which are arranging N and S magnet poles alternately. The prototype permanent magnet generator is tested with rotational speed and load condition. The experimental results have been shown that the coreless generator can generate It was found that 3D-printed coreless generator can generate electric energy up to 10300 mW for rotor with eight permanent magnets and 9000 mW for rotor with four permanent magnet at 5000 rpm, respectively.

1. Introduction

Permanent magnets generators (PMGs) exhibit a prospect to build energy efficient machine [1]. The stator and rotor are the major parts of PMGs. The main function of the stator is to generate alternating current from conductor windings while the rotor consists of permanent magnets as external magnetic field to generate the magnetic interaction with the stator. Nowadays PMGs are rapidly used in numerous applications in term of powering small or large-scale electronic systems, such as wind turbine application [2], hydroelectric generation [3] and others. Even though PMGs are able to achieve high efficiency compared with other generator types, however PMGs still more costly compared to other generators [4]. Therefore, many research works focus on the design of PMGs with basic requirements such as high output power, lightweight, simple construction, and low cost [1]. Unfortunately, the design consideration of those basic requirements should be carefully conducted based on the generator configuration and materials [4].

3D printing technology has become significant importance as rapid process to develop prototypes for various type of applications such as RF/microwave components [5], aerial vehicle wing structures [6], electromagnetic shielding [7] and biomedical applications [8]. Since the 3D printing technology has capabilities to develop complex geometries from the digital drawing, this technology becomes the major advantage in the manufacturing systems. The other advantage is that the 3D printing technology can be used to develop functionalities with other manufacturing methods [9]. Additionally, 3D printing technology can improve the design and lower the weight of the structure, resulting to a very cost-effective solution and reduction of lead times [10]. Based on these results, it is a worthy probability to use 3D printing technology for developing a generator. The development of generator using 3D-printing technology have been proposed by many researches works such as 3D-printed electrostatic generator [11], 3D-printed triboelectric generator [12], and 3D-printed materials low speed PM generator [13].

In this work, the possibilities of implementing 3D printed coreless generator with two kind of rotors for energy harvesting applications is reported. Following sections will explain the fabrication of the generator and the test results of the proposed 3D-printed coreless generator for both rotors.

Content from this work may be used under the terms of the Creative Commons Attribution 3.0 licence. Any further distribution of this work must maintain attribution to the author(s) and the title of the work, journal citation and DOI.
Published under licence by IOP Publishing Ltd

2. Materials and methods

2.1. Fabrication of 3D printed coreless generator

The generator is meant to convert mechanical energy into electrical energy. However, the coreless generator is a generator that use the core for coil of winding copper wire by non-ferrous materials. Most of these type of generators used Permanent Magnet (PM) in the rotor. PM in generator supplied the magnetic excitation as conversion device producing all electromagnetic energy. The advantage of coreless generator is that it can eliminate the strong interaction in magnetic field between PM and ferrous core resulting to a high resistant on rotating rotor. Therefore, the coreless generators are suitable for harnessing renewable energy such as wind, rain water, river water, and tidal energy, which produce relatively small energy. Hence, the coreless generator can capture a small amount of energy by smooth rotation without any resistant of magnetic field.

The main parts of the generator are stator and rotor. There are two types of generator: the first type has stator which is fixed at the middle and a rotor that is located at the outer stator housing, and the other type has rotor at the middle and the stator is fixed at the outer housing. The stator functioned as conductor winding to generate alternative current. While, the rotor, where the PM took place, is operating as external magnetic field to generate the magnetic interaction with the stator, thus can produce the alternative current.

All parts of the coreless generator are designed using Autodesk Inventor. In the stator design, there are four stacks for winding coil of copper wire that are placed perpendicular to the next coil. Two types of rotor are designed for this study. The four permanent magnets and eight permanent magnets, which both of rotors have slots for easier placement. The stator is the stationary part in the outer housing whereas rotor is the inner part which rotates. The specification of designed coreless generator are summarised in table 1.

Table 1. 3D printed coreless generator specification

Parameters	**Value**
Pole numbers	4 poles (rotor 1) and 8 poles (rotor 2)
Stator Slot for winding	4 slots
Rotor diameter	45 mm
Length of stator stack	39 mm
Stator inner diameter	50 mm
Stator outer diameter	97 mm
Air gab length	2,5 mm
Magnet dimension	60 x 10 x 10 mm^3

Anet A8 3D printer is used for the coreless generator fabrication with 1.75 mm of diameter of Black Polylactic Acid (PLA) as a 3D printing filament. Before running the 3D printer, the design of stator and rotor should be converted to stl file, then the file is opened in Cura software for printer setting. The setting are; the heating bed temperature and nozzle temperature set to 60°C and 210°C, respectively. The printing speed is 60 mm/s with the density of 60%. The support placement was chosen as "Everywhere" type to support hanging profile and can be removed easily without effecting the whole product. Figure 1 shows the construction of stator during operation of 3D printer.

A 0.60 mm copper conductor wire is used for winding coil for four slots of the stator with 500 turns in each slot as presented in figure 1. It is essential to count the turns for winding correctly and carefully. One turn of copper wire is set next to the other with enough tense and less gaps in between the turns of copper wire [4]. This is done to produce successful coils. At the rotors, four poles and eight poles neodymium magnets are chosen due to stronger magnets as Permanent Magnet (PM) with less volume to produce the external magnetic fields as shown in figure 2(a) and figure 2(b). The arrangement of the N and S magnet poles are set alternately.

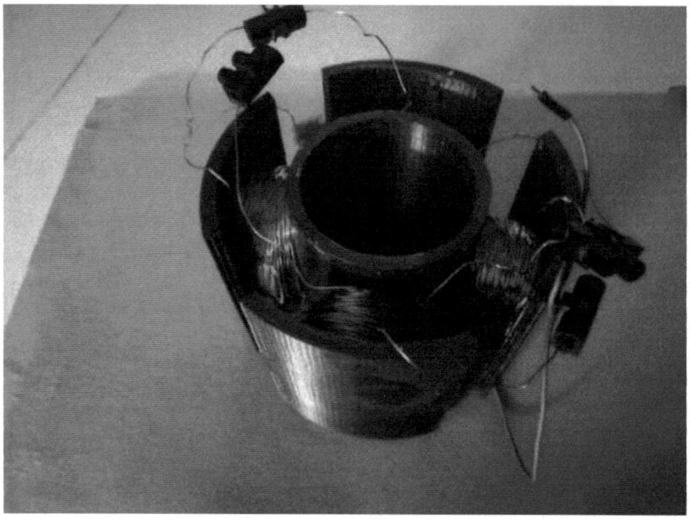

Figure 1. Stator with winding copper wire

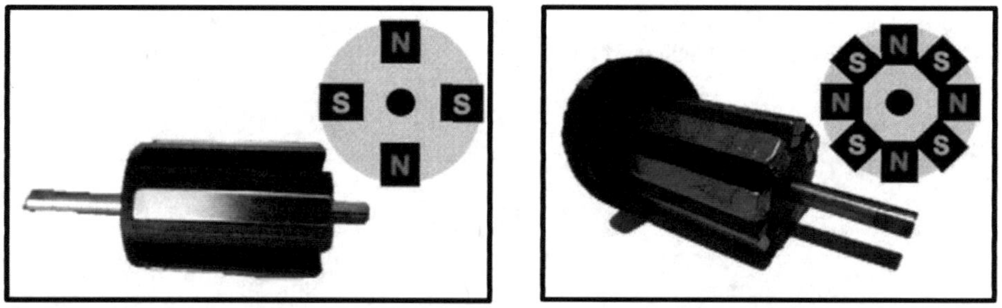

(a) (b)

Figure 2. (a) Rotor with four poles permanent magnet (b) Rotor with eight poles permanent magnet and their arrangements

2.2. Experimental setup

In conducting this prototype, there are four types of connection wire between the coils, due to the position of poles between 4 poles PM and 8 poles RM facing to the 4 winding coil stack in the stator are different. To get better understanding of winding coil, the illustration of winding coil is being labelling into inlet wire and outlet wire as shown in figure 3. Next, each of winding coils were connected with four different connections such as inlet to inlet, inlet to outlet, outlet to outlet, and others as summarized in figure 4.

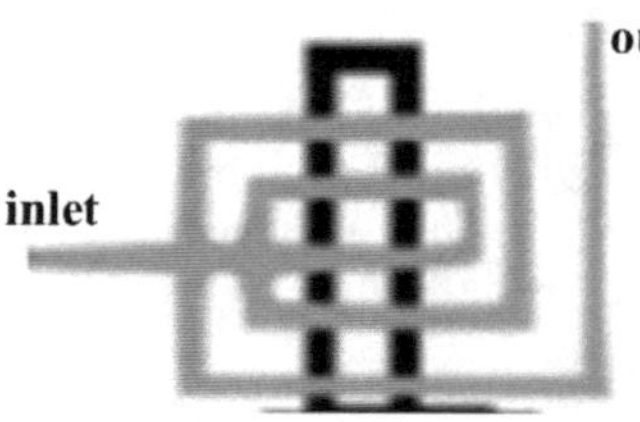

Figure 3. Labelling inlet and outlet at winding coil in stator

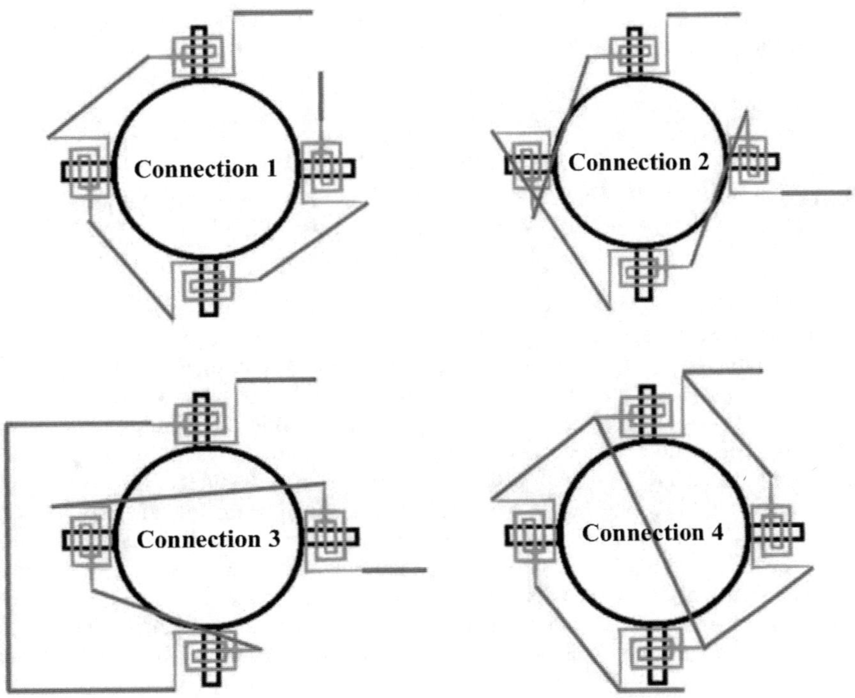

Figure 4. Four types of connection in winding coils in stator

The coreless generator was tested using 4 poles PM and 8 poles PM rotor for each four types of connection winding coil as shown in figure 4. Then the output of coreless generator is connected to a bridge rectifier to convert AC to DC current, load of 12 V bulb, voltmeter, and amp meter. The schematic diagram of experimental setup can be seen in figure 5. The prototype of the coreless generator was tested with rotational speed of 3000 rpm to observe the potential power output from the different connection winding coils. From these tests, two highest power output would be selected to further experiment.

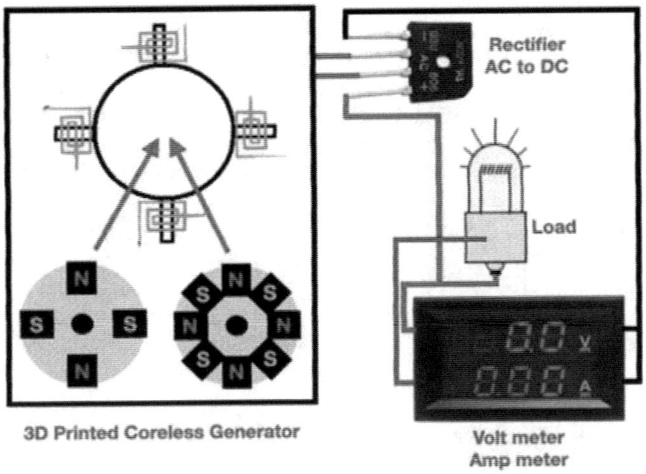

Figure 5. Schematic diagram of experimental setup

The further experiments were conducted with three different conditions. First, the coreless generator was running without any load to observe an amount of voltage and frequency of output generator that could be produced, which the frequency can be determined by speed of generator multiple with total number of poles and divided with 120.

Second, the coreless generator was running with 1 load of 12 V 5 W bulb. Lastly, the coreless generator was running with 2 loads of 12 V 5 W connected parallelly. For second and third experiments, the current produced from the coreless can be observed by Amp meter and Voltmeter for monitoring voltage. The coreless generator was connected to 100 W AC motor to drive the rotor during the experiments that can be controlled by speed controller from 0 rpm up to 6000 rpm.

3. Results and Discussion

Table 2 summarized the voltage and current produced from each rotors and winding coils connection of the coreless generator at speed of 3000 rpm. It can be observed that there is only one connection for each rotor that have power output, and the rest are not show any power output during these tests. The 4 poles PM rotor (rotor 1) is appropriate to the connection 2, while the 8 poles PM rotor (rotor 2) is appropriate to the connection 1. The connection 3 and connection 4 for both of rotors show no voltage and current are produced, therefore both of these connections are not suitable for rotor 1 and rotor 2.

As shown in table 2, the connection between each winding coil using inlet to outlet connection seems not properly work for 4 poles PM (rotor 1) when first stack of winding coil is facing N pole PM, while second stack of winding coil is facing S pole PM. At the same time, third stack of winding coil is facing N pole PM and fourth stack of winding is facing S pole PM. Meanwhile, if the connection between each winding coil using inlet to inlet, then outlet to outlet, the generator can generate voltage and current for rotor 1. Hence, this connection (2) is working well with rotor 1 arrangement that can generate power from coreless generator.

For rotor 2, the connection between each winding coil using inlet to outlet connection is properly working when all the stacks of winding coil are facing N pole PM. It means that the connection 1 is working when all the stacks of winding coil are facing the same pole (N or S) at the same time. Meanwhile, if the connection between each winding coil using inlet to inlet, then outlet to outlet, (connection 2) there no voltage and current are produced, since this connection can only work if the stacks of winding coil are facing N pole PM for odd number and S pole PM for even number. Hence, this connection (1) is working well with rotor 2 arrangement that can generate power from coreless generator.

The connection 2 for rotor 1 and the connection 1 for rotor 2 were chosen for further experiments. Figure 6 is summarized the voltage produced from the coreless generator for rotor 1 and rotor 2. It can be observed for both rotors that the increasing voltage is depending on the increasing of speed linearly. The generator with rotor 1 can generate voltage up to almost 25 volt at 5000 rpm with frequency of 165 Hz, while generator with rotor 2 can generated voltage up to 35 volt at 5000 rpm with frequency of 340 Hz. From this experiment, it can be concluded that this coreless generator for both rotors need higher speed to get higher voltage. Nevertheless, the generator with rotor 2 produces higher voltage and frequency than rotor 1. It means that the number of poles in rotor has a significant effect in producing voltage, where increasing poles from 4 poles to 8 poles can increase around 40% higher voltage at 5000 rpm. To implement to the renewable source of energy, the gear ratio between generator and turbine is crucial to obtain higher voltage.

By applying loads in the coreless generator with both rotors, the voltage produced reduces. Higher load is implemented, the lower voltage will be generated as shown in figure 7. However, the increasing load can increase the output current. At 4500 rpm, the voltage without load can generate almost 22 volt for generator with rotor 1, and 30 volts for generator with rotor 2. But then by giving 1 load and 2 loads, for generator with rotor 1, the voltage reduces to 16 volt and 10 volt, respectively. While the output current by giving 1 load and 2 loads, can increase from 350 mA to 500 mA at 6000 rpm.

Figure 8 shows the power out produced from the coreless generator with rotor 1 and rotor 2 for 1 load and 2 loads. It is clearly shown that both of load can produce similar results for each rotor. However, for rotor 2, power output from generator with 2 loads is slightly higher than with 1 load. Even thought, the voltage and current produced from 1 load and 2 loads are different for each rotor. This coreless

IOP Conf. Series: Materials Science and Engineering 1082 (2021) 012009 doi:10.1088/1757-899X/1082/1/012009

generator can produce up to 10300 mW and 9000 mW at 5000 rpm for rotor 2 and rotor 1, respectively. Therefore, it can be concluded that rotor 2 has higher power output than rotor 1 and any load applied on this coreless generator, the power will produce in similar output for each rotor.

Table 2. Voltage and current produced from rotor 1 and rotor 2 for each connection (1, 2, 3, 4) of coreless generator

Rotor 1 (4 poles PM)			**Rotor 2 (8 poles PM)**		
Connection of winding coil	**Voltage (V)**	**Current (A)**	**Connection of winding coil**	**Voltage (V)**	**Current (A)**
1 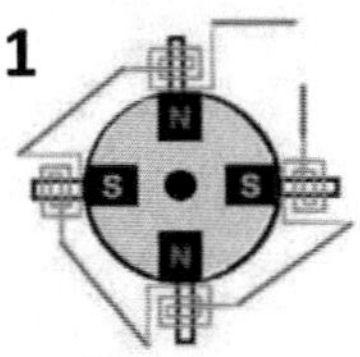	0	0	1 	14	0.32
2 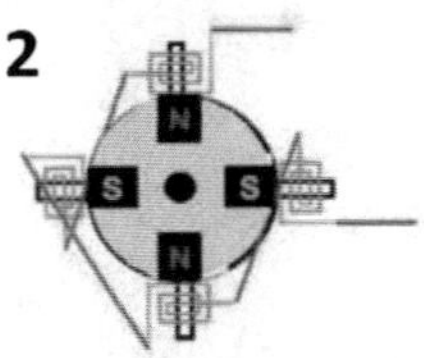	10	0,25	2 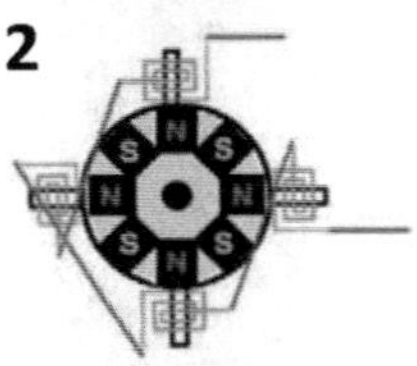	0	0
3 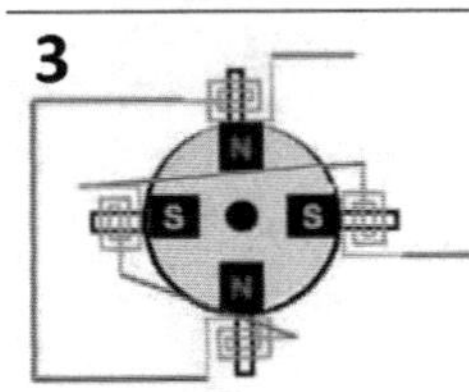	0	0	3 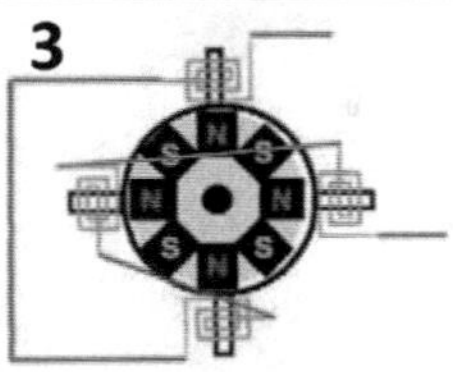	0	0
4 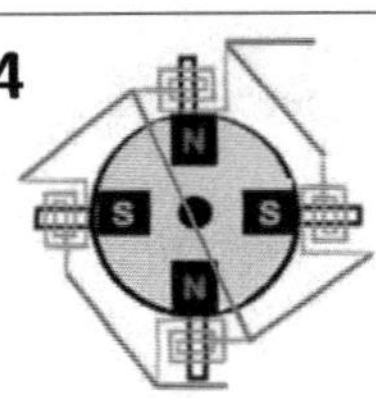	0	0	4 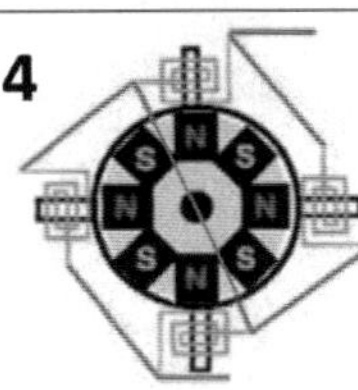	0	0

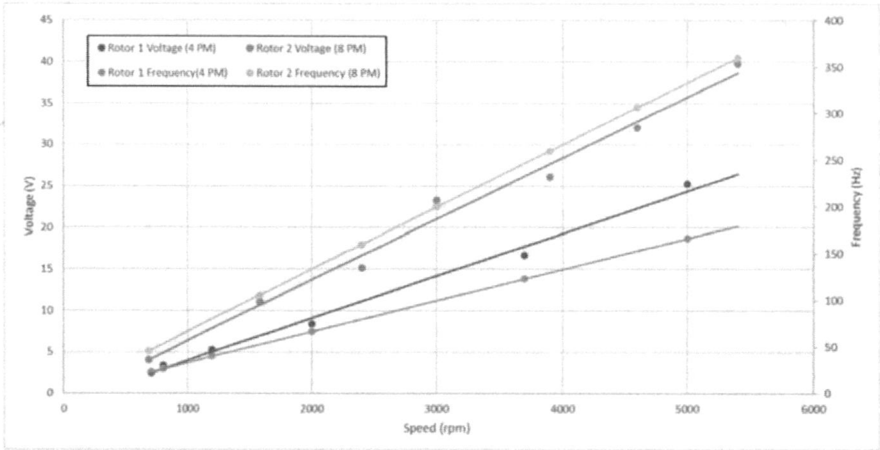

Figure 6. 3D printed coreless generator voltage and frequency with increasing speed for rotor 1 and rotor 2

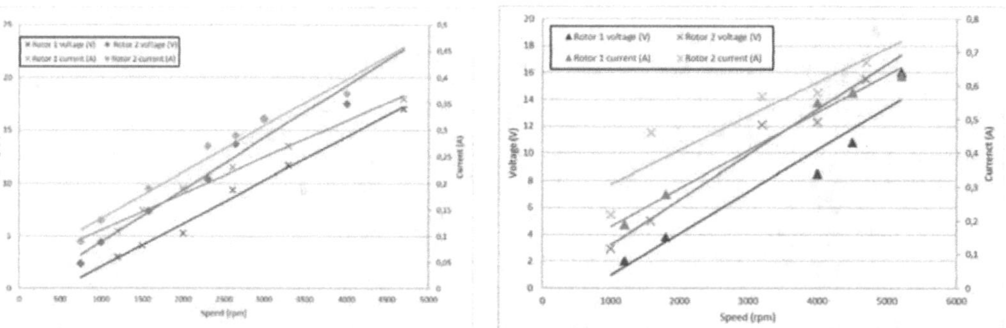

(a) (b)

Figure 7. 3D printed coreless generator voltage and current with increasing speed for rotor 1 and rotor 2 at 1 load (a) and 2 loads (b)

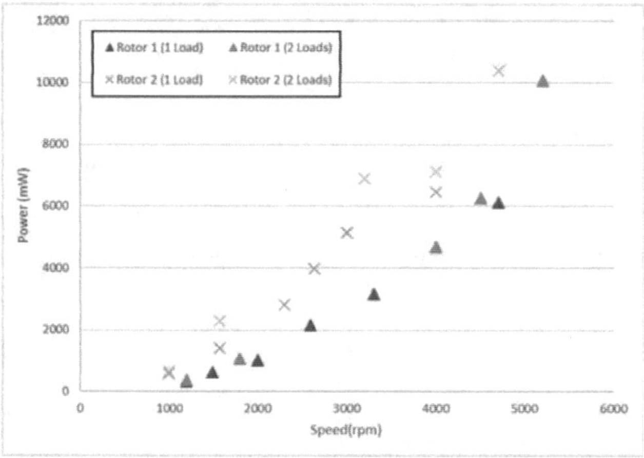

Figure 8. 3D printed coreless generator power with increasing speed for rotor 1 and rotor 2 with 1 load and 2 loads

4. Conclusions

In this work, the experiments with prototype of coreless generator created using 3D-printed are presented with different number of poles PM. The coreless generator is aimed to be applied in energy harvesting system. It was found that 3D-printed coreless generator can generate electric energy up to 10300 mW for rotor 2 and 9000 mW for rotor 1 at 5000 rpm that can be used in low-power system. Therefore, the development of coreless generator using 3D-printed exhibits promising potential to be used for this generator. Nevertheless, further studies are required to study the process developments in variety of filaments and setting on 3D printer.

Acknowledgements

This work is supported by Research and Technology Transfer Office, Bina Nusantara University as a part of Bina Nusantara University's International Research Grant entitled Development of Siphonic-Roof-Top Water Turbine for Green Technology with contract number: No.026/VR.RTT/IV/2020 and contract date: 6 April 2020.

References

[1] Kallaste A, Vaimann T and Pabut O 2012 Slow-speed ring-shaped permanent magnet generator for wind applications In *11th Int. Symp. Typical Problems in the Field of Electrical and Power Engineering* 66-9

[2] Soderlund L, Eriksson J T, Salonen J, Vihriala H and Perala R 1996 A permanent-magnet generator for wind power applications *IEEE Transactions on Magnetics* **32(4)** 2389-92

[3] Huzlik R, Ondrusek C and Palsovics N 2016 December. Design of permanent magnet generators for small hydro power plant In *2016 17th International Conference on Mechatronics-Mechatronika (ME)* 1-5 IEEE

[4] Latoufis K C, Messinis G M, Kotsampopoulos P C and Hatziargyriou N D 2012 Axial flux permanent magnet generator design for low cost manufacturing of small wind turbines *Wind engineering* **36(4)** 411-31

[5] Garcia C R, Rumpf R C, Tsang H H and Barton J H 2013 Effects of extreme surface roughness on 3D printed horn antenna *Electronics letters* **49(12)** 734-6

[6] Moon S K, Tan Y E, Hwang J and Yoon Y J 2014 Application of 3D printing technology for designing light-weight unmanned aerial vehicle wing structures *International Journal of Precision Engineering and Manufacturing-Green Technology* **1(3)** 223-8

[7] Viskadourakis Z, Vasilopoulos K C, Economou E N, Soukoulis C M and Kenanakis G 2017 Electromagnetic shielding effectiveness of 3D printed polymer composites *Applied Physics A* **123(12)** 736

[8] Wang K, Wu C, Qian Z, Zhang C, Wang B and Vannan M A 2016 Dual-material 3D printed metamaterials with tunable mechanical properties for patient-specific tissue-mimicking phantoms *Additive Manufacturing* **12** 31-7

[9] Haque R I, Farine P A and Briand D 2017 3D printed materials based triboelectric device for energy harvesting and sensing In *Multidisciplinary Digital Publishing Institute Proceedings* **1(4)** 580

[10] Mueller B 2012 Additive manufacturing technologies–Rapid prototyping to direct digital manufacturing *Assembly Automation*

[11] de Queiroz A C M and de Oliveira Filho L C M 2016 Energy harvesting with 3D-printed electrostatic generators In *2016 IEEE 7th Latin American Symposium on Circuits & Systems (LASCAS)* 127-30 IEEE

[12] Quan S, Bu L, Zhou Z, Xu W, Liang B and Song L 2016 3D-printed novel triboelectric generator based on saw-toothed button structure In *2016 13th IEEE International Conference on Solid-State and Integrated Circuit Technology (ICSICT)* 263-5 IEEE

[13] Soemphol C and Angkawisittpan N 2020 3D-printed materials based low-speed permanent magnet generator for energy harvesting applications *Materials Today: Proceedings* **22** 180-4

The integrated lean and green manufacturing system: a case study at the peeled loaf production

Ilyas[1], Suhendrianto[1], Raihan Dara Lufika[1], Syafira Meutia[1]
[1]Industrial Engineering Department, Universitas Syiah Kuala, Darussalam, Banda Aceh, Indonesia

E-mail: raihandlufika@unsyiah.ac.id

Abstract. The bakery industries soared as bread become one of the staple food for many people. UD. ABC is one of the bakeries in Banda Aceh that produces various types of bread and jam includes the peeled loaf. Peeled loaf uses many machines for its production process thus will produce waste and consume lots of energy. This study aims to analyze waste and measure energy use in the production process of peeled loaf based on the integrated Lean and Green Manufacturing concept. There are VSM and EVSM as a tool to describe the flow of material and energy usage from the production process. The activities of Value Added (VA), Non-Value Added (NVA) and Necessary Non-Value Added (NNVA) will be analyzed using PAM. The tool for improvement recommendations is five whys analysis. Then, the Future State Map (FSM) shows the improved design with a reduction in the lead time of 20.28 minutes. From this study also obtained the results of saving energy use as much as 0.13 kWh. This energy-saving will realize a lean and environmentally friendly production process at UD. ABC.

1. Introduction

There is a common requirement for today's companies to increase their work system's efficiency and their production system for the overall productivity of their company. The company can perform several actions in term of working and production systems to enhance efficiency, such as the use of raw materials, equipment, space and also the amount of labour required. It is necessary to control and supervise the whole process in the company to ensure all these aspects work accordingly and environmentally friendly without waste. Waste defined as an activity in a process which adds cost and time but doesn't add value to a product/service from a consumer point of view.

Waste is the focus major on the lean manufacturing system[1] as its goal is to have a waste-free production. *Muda* is a Japanese word that means waste [2] which is related to lean manufacturing system. The idea is to make improvements on things that cause waste and are related to the environment. The industry should be aware of the issues of global warming, natural resource scarcity, and waste management. The manufacturing industry is an industry that consumes a lot of energy in the production process. However, many companies now concern about the environment and trying to make an environmentally friendly industry (Green Manufacturing). Green manufacturing mostly focuses on the philosophy area instead of standard or process [3]. The whole idea of green thinking is to understand that nature is limited [4]. In addition, the word "Green" is a verb to portray a process as an activity to reduce the environmental impact of the manufacturing process or system. Energy

Content from this work may be used under the terms of the Creative Commons Attribution 3.0 licence. Any further distribution of this work must maintain attribution to the author(s) and the title of the work, journal citation and DOI.
Published under licence by IOP Publishing Ltd

consumption is one focus on green manufacturing with the philosophy of environmental impact as the prior issue.

UD. ABC is one of the bakeries in Banda Aceh that produces various types of bread and jam includes the peeled loaf. There are varieties of machines and equipment on the production floor which requires lots of energy such as electricity. In term of the peeled loaf, the production needs as much as six machines from the whole production process activities.

The whole production of the peeled loaf which starts from raw materials process until it becomes the final product requires electricity. Therefore, it takes a lot of electrical energy in one production process. Based on the observations, the waste on the production process mostly happened during the waiting time and defects. The defect causes energy waste in the production process as well as the raw material consumption. It leads to company loss in term of financial and productivity because the company has to pay the additional costs and time for managing waste.

Therefore, the research aims to solve the problem at the production system of the peeled loaf using the integration of Lean and Green Manufacturing. This study consists of steps such as identification of waste, identification of activities, mapping the state map and energy consumption, analysis, and future suggestion.

2. Methodology

2.1. Data Collection

The data collection consist of two stages which are collecting the primary and secondary data. The primary data consists of process sequence data production, cycle time data and engine power. Meanwhile, secondary data consists of data number of production, number of workers and hours work.

2.2. Data Processing

Data processing begins with testing data uniformity and adequacy for cycle time. Then, we calculate the standard time for the Current State Map (CSM) design and the energy consumption for the Energy Value Stream Map (EVSM). The next step is to analysis the CSM and EVSM for further waste identification using PAM. After that, the waste analysis is using the five whys tools. The last step is to design the Future State Map (FSM) to presents alternative solutions to the problem.

3. Results and discussion

3.1. Current State Map making

The Current State Map (CSM) refers to the calculation of standard time that consists of tests data uniformity, data sufficiency test, normal time calculation and standard time calculation.

3.1.1. Standard time calculation

The data uniformity test including calculate the average cycle time, standard deviation, Limit Upper Class (BKA) and Lower Class Limits (BKB) and create control maps for activities of the production process of the peeled loaf. If the data is enough, then we conduct a data sufficiency test. Data sufficiency test using the confidence level at 95% ($k = 2$) and an error rate of 5%. Table 1 is a summary of the adequacy test data. Normal time calculation is using an adjustment factor based on westing house tables and cycle times. Table 2 is a recapitulation of time calculations normal. Standard time is using the allowance factor and value normal time. Table 3 shows the recapitulation of standard time.

3.2. Current State Map

Figure 1 represents the CSM of the process production of peeled plain bread. The CSM shows that the longest process of the whole peeled loaf production is a process of forming/rolling and placing of dough that is for 99.38 minutes. The fastest processing time is the time for moving the dough to the table dough that is 0.50 minutes. Lead Time from the production process of the peeled loaf is 466.89 minutes with the value-added time of 292.80 minutes.

Table 1. Recapitulation of data adequacy test.

Activities	N'	N	Information
Mixing and stirring of raw materials	1,52	5	Enough
Weighing and rounding dough	3,69		Enough
Forming / rolling and placing of dough	0,31		Enough
Baking dough	0,03		Enough
Loaf cooling	1,01		Enough
Peeling loaf	2,85		Enough
Loaf cutting	4,04		Enough
Loaf packaging	2,96		Enough

Table 2. Recapitulation of normal time calculation.

No	Activities	Ws	Adjustment	Wn
1	Mixing and stirring of raw materials	20,60	1,11	22,87
2	Weighing and rounding dough	8,67	1,05	9,11
3	Forming / rolling and placing of dough	75,84	1,12	84,94
4	Baking dough	50,49	1,12	56,55
5	Loaf cooling	60,49	1,11	67,15
6	Peeling loaf	13,39	1,05	14,06
7	Loaf cutting	39,72	1,05	41,70
8	Loaf packaging	32,78	1	32,78

Table 3. Recapitulation of standard time calculation.

No	Activities	Wn	Leniency Factor	Wb
1	Mixing and stirring of raw materials	22,87	0,17	26,64
2	Weighing and rounding dough	9,11	0,16	10,57
3	Forming / rolling and placing of dough	84,94	0,17	99,38
4	Baking dough	56,55	0,21	68,14
5	Loaf cooling	67,15	0,12	75,21
6	Peeling loaf	14,06	0,17	16,45
7	Loaf cutting	41,70	0,14	47,54
8	Loaf packaging	32,78	0,14	37,37

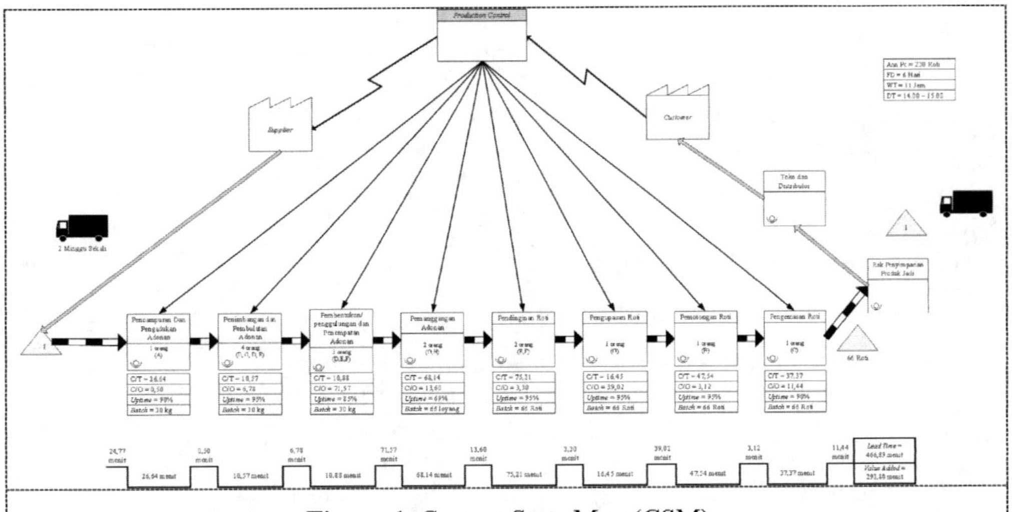

Figure 1. Current State Map (CSM).

3.3. Energy Value Stream Mapping (EVSM)

The calculation of energy consumption for Value Added (VA), Non-Value Added (NVA) activities and Necessary Non-Value Added (NNVA) is the basic for EVSM design (Figure 2). Table 4 is a recapitulation of energy consumption in the process production of the peeled loaf.

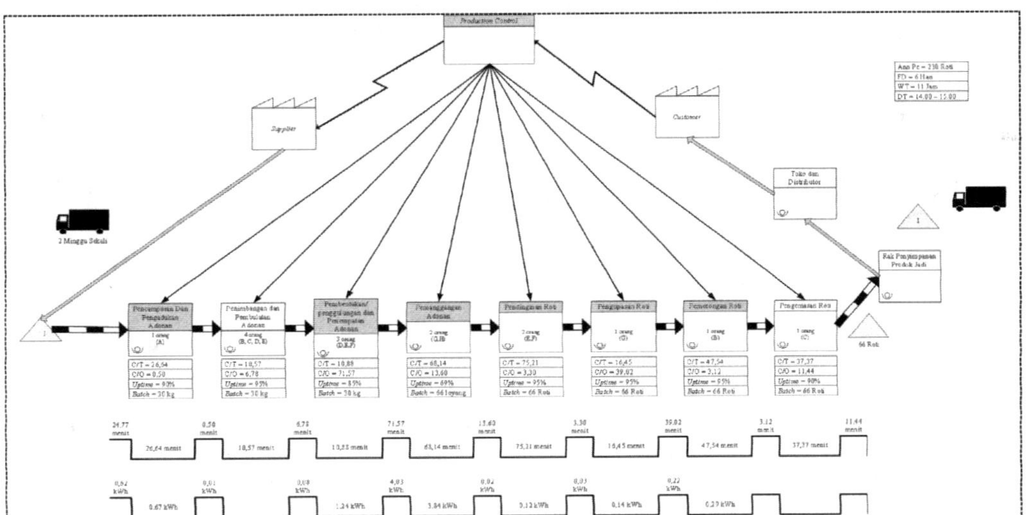

Figure 2. Energy Value Stream Map (EVSM).

IOP Conf. Series: Materials Science and Engineering 1082 (2021) 012010 doi:10.1088/1757-899X/1082/1/012010

Table 4. Recapitulation of energy consumption in the process production of the peeled loaf.

No	Activities	Machine	Energy usage (kWh)	Comment
1	Carry dough ingredients from the warehouse to mixing station	*Mixer*	0,59	NNVA
2	Raw materials waiting to be mixed	*Mixer*	0,03	NVA
3	Mixing and stirring the dough	*Mixer*	0,67	VA
4	Transfer the dough to the dough table	*Mixer*	0,01	NNVA
5	Weighing and rounding the dough	*Moulding Machine* NFZ-380	0,13	VA
6	Transfer the dough to a rolling machine	*Moulding Machine* NFZ-380	0,04	NNVA
7	Take a baking sheet	*Moulding Machine* NFZ-380	0,04	NNVA
8	Forming/rolling and placing the dough	*Moulding Machine* NFZ-380	1,24	VA
9	Arrange a baking sheet	*Oven*	0,19	NNVA
10	Waiting for the dough to rise	*Oven*	3,47	NNVA
11	Bring the dough to the grill	*Oven*	0,37	NNVA
12	Baking dough	*Oven*	3,84	VA
13	Transferring and removing loaves from the pan to the cooling table	Fan	0,02	NNVA
14	Arrange the loaf to be cooled	Fan	0,01	NNVA
15	Cooling loaf	Fan	0,12	VA
16	Move loaf to a stripping station	Bread skin peeling machine	0,03	NNVA
17	Stripping loaf	Bread skin peeling machine	0,14	VA
18	Arrange the loaf on a rack			
19	Waiting for the loaf to be cut	Bread cutting machine	0,07	NNVA
20	Waiting for the loaf to be cut by the worker	Bread cutting machine	0,13	NVA
21	Transferring the loaf to the chopping machine	Bread cutting machine	0,02	NNVA
22	Cut the loaf	Bread cutting machine	0,29	VA
23	Transferring loaf to the packaging department			
24	Do the packaging			
25	Transfer the loaf to the box			
26	The loaf is moved to the distribution place			
Total			11,45	

3.4. Activity classification using Process Activity Mapping (PAM)

The tools for waste identification is PAM. The first step for PAM design is the classification of activities to three categories, namely Value Added (VA), Non-Value Added (NVA) and Necessary Non-Value Added (NNVA). The categories in the production process are the result of observations and interviews with four people in the industry which consists of owner, manager and two workers. Table 5 is a summary of PAM. Table 5 shows that the two highest time-consuming activities are operations and transportation with the total times are 299.47 minutes and 61.84 minutes respectively. Figure 3 is a comparison of activities VA, NVA and NNVA in percentage form. The takt time for this problem is 165 minutes/cycle which means there is no problem in the production system.

Table 5. PAM Summary.

Category	Time (minutes)	Total Activities
Operation	299,47	10
Transportation	61,84	10
Inspection	0,00	0
Delay	95,21	4
Storage	10,38	2

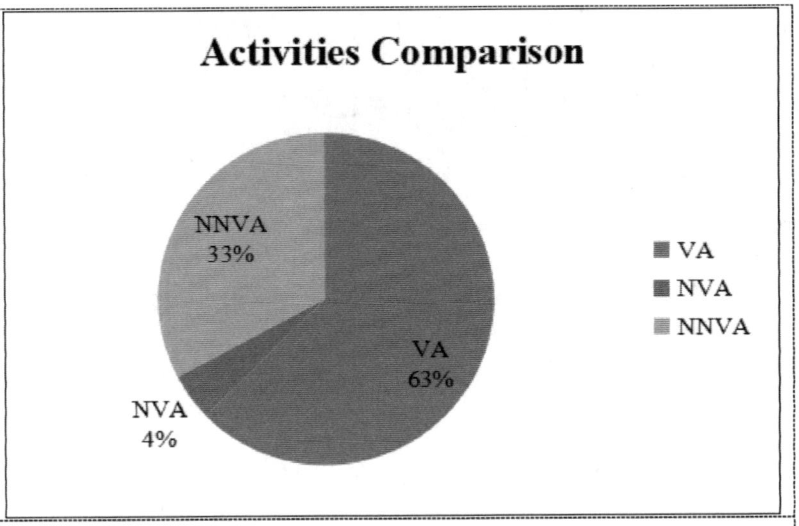

Figure 3. The comparison of VA, NVA, and NNVA activities.

3.5. Waste Analysis

The waste analysis is to identify waste by observing and interviewing the stakeholders on the industry. There are three types of waste rank, namely high, medium and low. The rank is to determine the priority for improving waste. Table 6 is the rank of seven wastes, with five analysis, and corrective action plan (as shown in Table 7 and Table 8).

Table 6. Seven wastes rank.

Waste Types	High	Medium	Low	Object
Defects			v	Loaf that is burnt, deflated or wrongly cut
Excessive transportation		v		Raw material transfer activities
Inappropriate processing		v		Repetitive weighing activities
Unnecessary motion			v	No excessive movement of workers or materials
Unnecessary inventory			v	Dough or bread is directly processed or distributed
Overproduction		v		There was bread returned
Waiting		v		Bread waiting to be cut by workers

Table 7. Five whys analysis

No	Waste Types	Waste	*Why*	*Why*	*Why*	*Why*	*Why*
1	*Excessive Transportation*	Raw material transportation time is too long	Long-distance between raw material place and mixing station	The factory layout is still not following the production flow			
2	*Inappropriate processing*	*Over processing*	There is a process with inefficient work	There is a process that takes too long	The process of weighing dough repeatedly	Workers are hesitant when weighing the dough that has been cut	Workers are not yet proficient at weighing dough
3	*Overproduction*	Excess production of peeled loaf	Production of the loaf using a push or a make-to -tock system	The company does not estimate demand			
4	*Waiting*	Waiting time in the process of mixing raw materials and cutting loaf	The process does not run continuously	There is a buffer in the loaf slicing place	Waiting for workers to finish from other work	There are more workers at the forming dough station	There is no specific duty for the workers

Table 8. Corrective action plan.

No	Waste	Root problem	Corrective Plan
1	*Excessive Transportation*	Factory layout that doesn't match the workflow	Move raw materials to a space closer to the mixing station
2	*Inappropriate Processing*	Unskilled workers	Provide instructions or weighing procedures by skilled workers or industry
3	*Overproduction*	The industry does not make demand forecasts or forecasts	Industry performs forecasts for demand or forecasting
4	*Waiting*	There is no specific duty for the workers	A regular work schedule is applied according to the needs on the production floor

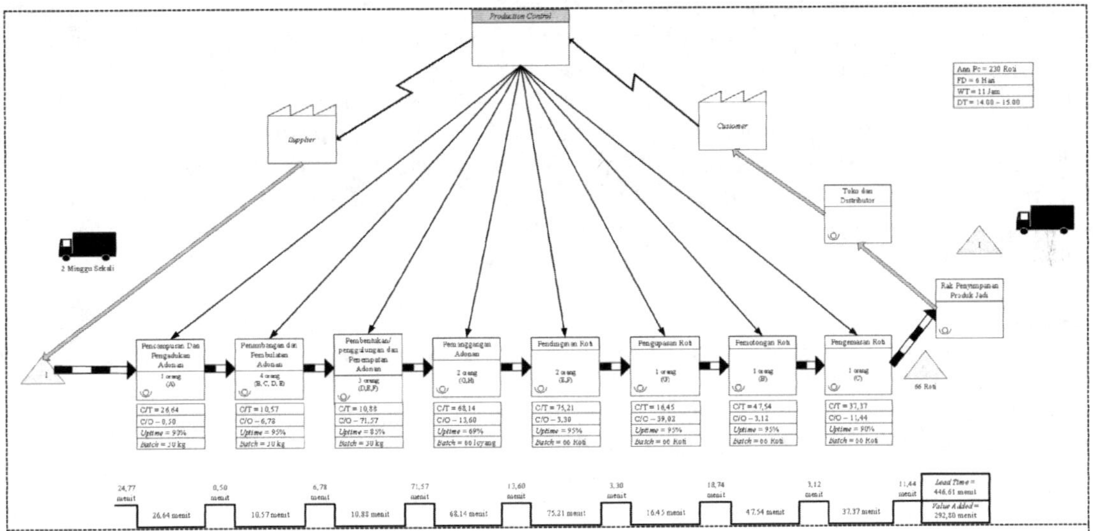

Figure 4. Future State Map (FSM).

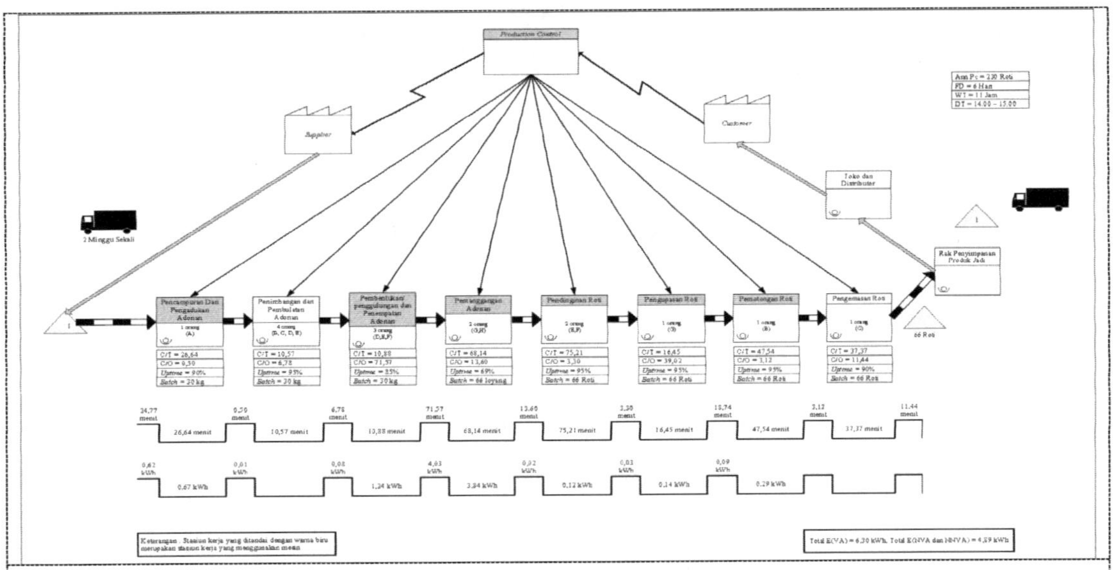

Figure 5. Future Energy Value Stream Map (FEVSM).

3.6. Future State Map (FSM)
Future State Map (FSM) refers to previous CSM and EVSM. Figure 4 is the FSM of the peeled loaf production process. It shows that there is a lead time reduction into 446.61 minutes. Figure 5 is Future Energy Value Stream Mapping of the peeled loaf process. There is a reduction in NVA energy consumption to 5.02 kWh. .

4. Summary
There are two future suggestion refers to CSM and FSM. There should be rule modifications for placing workers on multiple workstations and elimination of the waiting for loaf cutting activity. While there are four suggestions refers to EVSM and FEVSM. The propose solutions are transferring raw materials to the other room closer to the station of raw material mixing, providing signposts for turning off the engine, performing a Just-In-Time production system, and using recycled products for packaging.

In conclusion, there are four priority wastes in the peeled loaf production which are excessive transportation, inappropriate processing, overproduction and waiting. Then the lead time is reduced by 20.28 minutes if the company successfully apply the solutions as states in the FSM. The energy consumption for one cycle of the peeled loaf production process is 11.45 kWh which can be reduced by 0.13 kWh/cycle.

References
[1] Mostafa S, and Dumrak J, 2015, *Waste elimination for manufacturing sustainability*, Procedia Manufacturing, **2**, 11–16.
[2] Santos J, Wysk R A, and Torres M J, 2006, *Improving Production with Lean Thinking*, New Jersey, United States of America: John Wiley & Sons, Inc., Hoboken, New Jersey.
[3] Maruthi G D, and Rashmi R, 2015, *Green Manufacturing: It's Tools and Techniques that can be implemented in Manufacturing Sectors*, Materials Today: Proceedings, **2(4–5)**, 3350–3355.
[4] Found P, 2016, *Green Factory*, Taylor & Francis Group.
[5] Wahyudi, 2003, *Memproduksi roti*, Jakarta: Direktorat Pendidikan Menengah Kejuruan Direktorat Jenderal Pendidikan Dasar dan Menengah Departemen Pendidikan Nasional.
[6] Gaspersz V, and Fontana A, 2011, *Organizational excellence: systematic continuous improvement and innovation*, Bogor: Vinchristo Publication.
[7] Amaranti R, Irianto D, and Govindaraju R, 2017, *Green Manufacturing: Kajian Literatur*, 8–9.

Performance of GLCM Algorithm for Extracting Features to Differentiate Normal and Abnormal Brain Images

Zul Indra[1], Yessi Jusman[2,*]

[1]Department of Informatics Engineering, Faculty of Engineering, Abdurrab University, Kota Pekanbaru, Riau 28291, Indonesia

[2,*]Department of Electrical Engineering, Faculty of Engineering, Universitas Muhammadiyah Yogyakarta, Bantul, Daerah Istimewa Yogyakarta 55183, Indonesia

Corresponding author E-mail: yjusman@umy.ac.id

Abstract. Brain cancer is a malignant brain tumor that can spread quickly to other parts of the brain and spine. However, not all tumors are malignant and can be treated before they become malignant. The purpose of this study is to discover brain abnormalities based on CT scan images by using T-test algorithm. Thus, it can be one of solution for early detection of brain abnormalities in order to treat it before it becomes a malignant tumor or cancer. As dataset, this research using 40 images consisting of 20 normal brain images and 20 abnormal brain images. There are two algorithms which are used in this research i.e. Gray level co-occurrences matrix (GLCM) for feature extraction and T-Test for brain image classification. Prior to feature extraction, brain image is converted to *Graycomatrix* in order to adjust the brightness of the image. The final step is image classification by using the T-test algorithm. From 40 test results which are used in this study, GLCM method can extract 8 features that can significantly distinguish the image of normal brain and abnormal brain. For the T-test algorithm, it is found that each feature has a P-value <0.05 which means that extracted features can be used for the further classification process of brain image abnormality. Thus, it can be inferred that this research framework which is employed the GLCM and T-test algorithm can be used to assist the process of early diagnosis of brain cancer.

1. Introduction

Brain tumours is a of main public health problem which is one of the biggest causes of death in the world [1]. Due to level of mortality, brain tumour is the second dangerous disease that causes death for men aged 20-30 years and is the fifth dangerous disease that causes death for women aged 20-30 years. In 2010, International Agency for Research on Cancer declared that more than 126,000 people in the world each year have brain tumour and more than 97,000 people die [2]. In addition to having a very high mortality rate, brain cancer is also a very dangerous disease in terms of the difficulty to make an initial diagnosis [3]. In the early stages, the brain tumour is very difficult to be discovered because the tumour boundary is still unclear, the contrast is too low and sometimes resembles as a normal tissue [4].

Brain cancer is a malignant brain tumour that can spread rapidly to the other parts of the brain and spine [5]. However, not all brain tumours are malignant and can be categorized as cancer [6]. There are also benign brain tumours. Benign brain tumours are a group of brain cells that grow slowly and do not spread to the other parts [7]. The brain tumours itself is an abnormal and uncontrolled

Content from this work may be used under the terms of the Creative Commons Attribution 3.0 licence. Any further distribution of this work must maintain attribution to the author(s) and the title of the work, journal citation and DOI.
Published under licence by IOP Publishing Ltd

growth of brain cells. In the brain, tumour can develop from cells that make up brain tissue, from nerves that come in and out of the brain, and from the protective membranes of the brain and spinal cord. Brain cancer is an event where abnormal tissue growth occurs in the brain [5]. Abnormal tissue growth in the brain can be indicated by symptoms that appear, including vomiting, seizures, decreased nerve function, and emotional. In order to treat brain cancer effectively, there is a need for the right strategy to deal with it. One of strategy that can be done is to help the process of early diagnosis of brain tumours [8, 9].

It can be inferred that early diagnosis is very necessary in terms of handling brain tumour to prevent further damage and even death for the patient. Along with the rapid development of technology, one technique that is quite widely applied to help diagnose diseases is the digital image processing techniques. Digital image processing is utilized to analyse the medical images which are produced by CT scans or magnetic imaging resonance (MRI). However, those medical images are very high-resolution image so that it requires a lot of time to be examined by experts. To overcome this time-consuming problem, a feature selection method was presented in the digital image processing technique. Feature selection is a pre-processing step to reduce the complexity of the image during diagnostic models [9]. The main goal of the feature extraction is to discover the features which is the best representation and contain fewer parameter for medical images [10, 11]. By finding the right features, the process of diagnosing brain tumours can be performed faster because non-essential features in medical images of the brain have been successfully eliminated.

This research is intended to propose an intelligence system which can assist the early diagnosis of brain tumour detection by providing training data for further classification system. These training data are obtained from the feature extraction process of medical image by using GLCM algorithm. Gray level co-occurrences matrix (GLCM) is one most popular texture-based feature extraction algorithm which can assign the textural relationship between the pixels images. The main principle of the GLCM is to perform operations according to second-order statistics in the image [12, 13, 14]. The research framework for this study consists of two main stages namely features extraction of the brain image and the validation of these extracted features. A set of feature matrix which can represent the medical image with fewer parameters is the final output from this features extraction stage. After that, these feature matrixes are then validated by using T-Test algorithm and P-Value standard.

2. Related Work

The main challenge of feature extraction is to determine the appropriate features in order to reduce the complexity of medical images without reducing the quality of the information in medical images. Up to this time, there have been several studies that have been attempted to be carried out regarding features extraction of brain images. The summary of the related work for feature extraction of brain tumour is explained in Table 1.

From Table 1, it can be inferred that the research to find the best way in extracting brain image feature is still ongoing today. Furthermore, until now, there has not been one study that conducted research to apply data validation of the generated features. Therefore, this research proposed a new approach for extracting the brain images feature. This effort is hoped will increase the quality the feature extraction process as well as to assist the early diagnosis of brain tumour detection.

Table 1. Work Related to Feature Extraction of Brain Tumour Image

Author Name and Year	Feature Extraction Algorithm	Feature Reduction and Optimation	Feature Validation
Sharma et al. (2014) [15]	GLCM	No	No
Sudha et al (2014) [16]	GLCM and GLRM	No	No
Demirhan et al (2014) [17]	SWT	No	No
Gaikwad, S. B., & Joshi, M (2015) [18]	PCA	PCA	No
Vidyarthi, A., & Mittal, N (2015)	Gabor wavelet and DWT	No	No

[19]			
P. M. Sanjeev Kumar and S. Chatteijee (2016) [20]	GLCM	No	No
Banday, S. A., & Mir, A. H. (2016) [21]	GLCM and GLRLM	PCA	No
Mathew et al (2017) [22]	DWT and GLCM	PCA	No
Islam et al (2017) [23]	BWT and GLCM	PCA	No
Sergaki et al (2018) [24]	GLCM	No	No
Dubey et al (2018) [25]	GLCM	No	No
Gopika, G. S., & Rajasree, R. S. (2018) [26]	GLCM	PCA	No

3. Methodology

3.1. Research Framework

The research framework for this study is divided into two main stages namely features extraction of the brain image and validation of the extracted features. The flowchart for this research is illustrated in Figure 1.

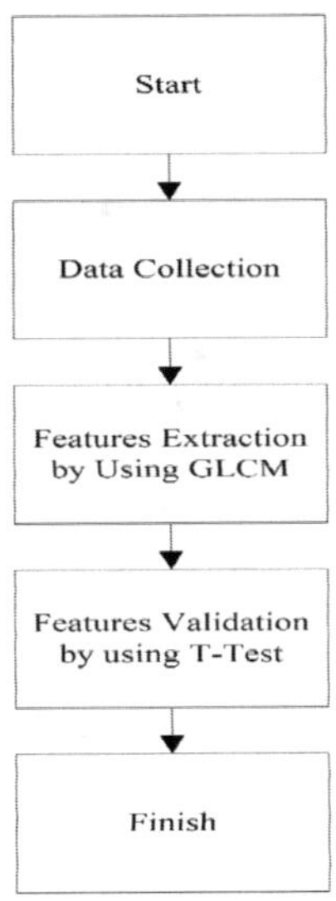

Figure 1. Research Framework.

3.2. Data Collection

As in general, data collection is the first section for this research. In this research, data collection is performed by collecting the brain images from Radiopedia , a wiki-based international collaborative radiology educational web resource. To conduct an experiment, this study uses two types of images, namely normal and abnormal brain images. The total dataset of this study is 20 brain images for each type of image. In addition, these used brain images is captured in 904 x 1024 pixels. Furthermore, each of images is then captured in ten times for different positions.

3.3. Feature Extraction

The next process after data collection is feature extraction as presented in Figure 2. Feature extraction is the process of taking the characteristics contained in an object of the image. However prior to the feature extraction section, the sampling image of the brain is converted into grayscale image to adjust the brightness of the image. Grey Level Co-Concurrent Matrix (GLCM) algorithm is employed to carry out this feature extraction. Feature extraction with texture analysis is aimed to obtain the characteristics of a grayscale image in order to distinguish one image from another image. In this study, the image is extracted based on four characteristics namely entropy parameters, contrast, energy, correlation. Each characteristic is an important feature because it contains valuable information in an image. These characteristics can be considered as the surface structure of an image.

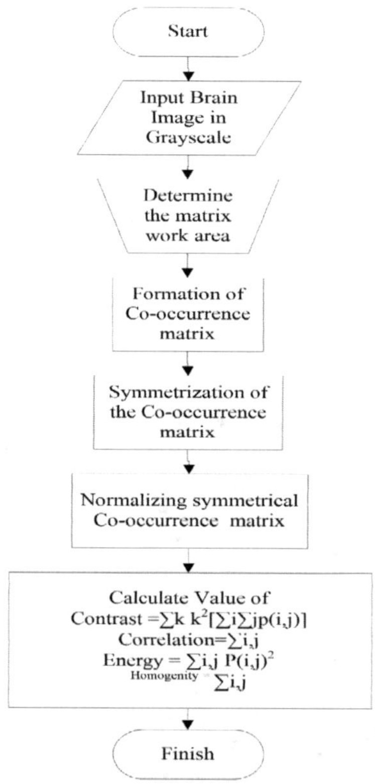

Figure 2. Flowchart of Feature Extraction.

3.4. Feature Extraction

The last section in this study is validating the extracted features as presented in Figure 3. This section is intended to validate the feature whether the feature can be used for the further classification process of brain image abnormality. This study utilizes the T-test algorithm to achieve this purpose [27,28,29,30]. Apart from that, the P-Value standard is used to carry out this feature validation process [31,32].

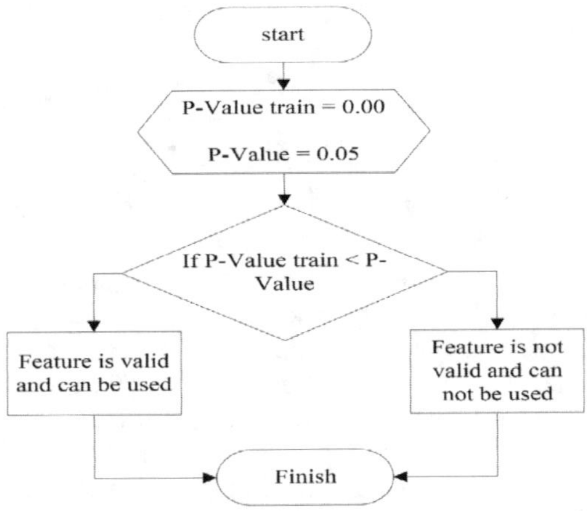

Figure 3. Flowchart of T-Test Validation.

4. Results & Analysis

4.1. Result of Feature Extraction

As explained before, there are two types of images which are used in this study i.e. normal and abnormal brain images. Both Normal and abnormal images have differences texture that can be seen by naked eye. This difference becomes a reference to determine the type of normal and abnormal images. The difference in terms of texture can be seen in the Figures 4 and 5.

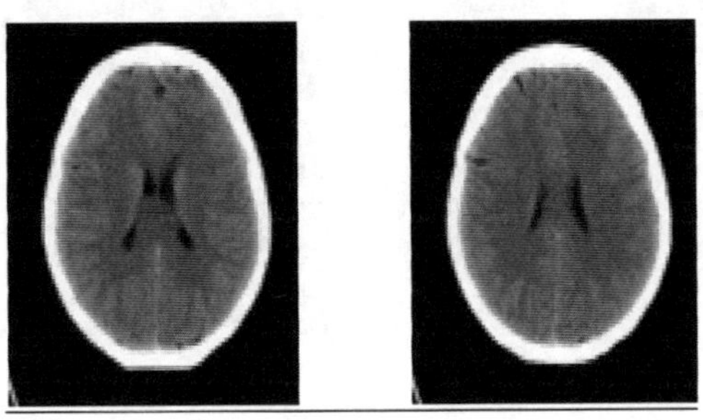

Figure 4. Normal Brain Image.

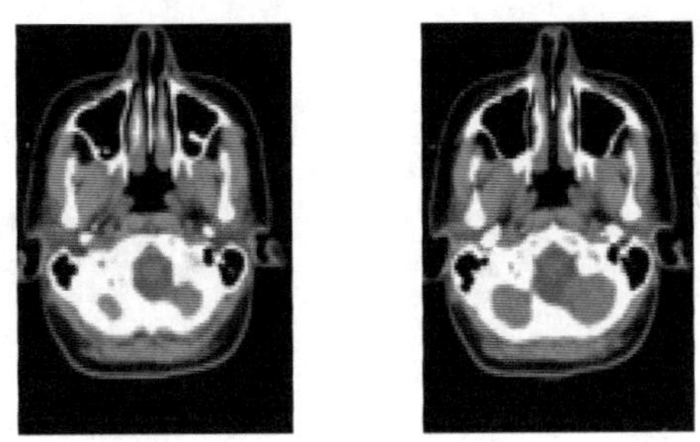

Figure 5. Abnormal Brain Image.

Prior to extract the feature, the brain images which are used as the dataset have to be converted into Graycomatrix form. This conversion process is intended so that the image has a better contrast value. Comparison of brain images in Grayscale form with brain images that have been changed to Graycomatrix. The value will make it easier for us to distinguish between normal brain cells and abnormal brain cells. Once the conversion process of the image into Graycomatrix is completed, the next step is to extract feature of the converted Graycomatrix images. As explained before, the feature of image is extracted based on four entropy characteristics namely contrast, energy, correlation and homogeneity.

Feature extraction can be varied according to the displacement and orientation value [33,34,35,36,37]. There are two displacement values that can be used namely d =1 and d = 2 and four orientation values θ = 0°, 45°, 90°, and 135° to form the GLCM matrix. The results of the feature extraction and its variations can be seen in Table 2.

Table 2. Extracted Feature

Angle Used	Name of Feature
4 Angles	Contrast1, Corellation1, Energy1, Homogeneity1
8 Angles	Contrast1, Contrast2, Corellation1, Corellation12, Energy1, Energy2, Homogeneity1, Homogeneity 2

From Table 2, it can be seen that there are several features produced from the feature extraction process according to the used of displacement values. When only d =1 displacement value is used, it will produce four features namely contrast1, corellation1, energy1, homogeneity1. However, if both of the two of displacement value (θ d =1 and d =2) are used, the GLCM algorithm will produced eight feature. The displacement value d =1 produces four features namely contrast1, correlation1, energy1 and homogeneity1. Four other features namely contrast2, correlation2, energy2 and homogeneity2 are obtained from the use of the displacement value d =2. Once the value of each feature characteristic is obtained, the feature extraction process is completely done. The next section is the validation of these extracted values.

4.2. Result of Feature Validation

The validation of the extracted features is the last section in this study which is intended to validate the feature whether the feature can be used for the classification process of brain image abnormality. The validation process is performed by using T-test algorithm (i.e. mean, standard deviation, and P values) The result of this validation process is described in the following Tables 3 and 4. From Tables 3 and 4, it is discovered that all features extracted using GLCM Algorithm have P-value lower than 0.05 where the value of 0.05 is the threshold of the features feasibility. Thus it can be concluded that all of the extracted features are proper for the brain cell classification.

Table 3. Feature Validation for 4 Angles

Feature	Image Type	Mean	St.Dev	P-Value
Contrast1	Abnormal	0,07	0,02	0,000
	Normal	0,11	0,02	
Correlation1	Abnormal	0,99	0,00	0,000
	Normal	0,99	0,00	
Energy1	Abnormal	0,34	0,03	0,000
	Normal	0,26	0,04	
Homogeneity1	Abnormal	0,96	0,00	0,000

Table 4. Feature Validation for 8 Angles

Feature	Image Type	Mean	St.Dev	P-Value
Contrast1	Abnormal	0,65	0,02	0,000
	Normal	0,10	0,02	
Contrast2	Abnormal	0,10	0,03	0,000
	Normal	0,17	0,04	
Correlation1	Abnormal	0,99	0,00	0,000
	Normal	0,99	0,00	
Correlation2	Abnormal	0,99	0,00	0,000
	Normal	0,99	0,00	
Energy1	Abnormal	0,34	0,03	0,000
	Normal	0,26	0,04	

Energy2	Abnormal	0,34	0,04	0,000
	Normal	0,26	0,06	
Homogeneity1	Abnormal	0,98	0,00	0,007
	Normal	0,97	0,00	
Homogeneity2	Abnormal	0,97	0,05	0,004
	Normal	0,96	0,01	

5. Conclusion

Based on the results of the research that has been conducted, it proved that the Gray level co-occurrences matrix (GLCM) method can be used to extract features to distinguish normal and abnormal brain images. The features extracted in this study are based of four characteristics i.e. contrast, correlation, energy and homogeneity. All features extracted using GLCM Algorithm have P-value lower than 0.0 where the value of 0.05. It can be inferred that this research framework which is employed the Gray level co-occurrences matrix (GLCM) and T-test algorithm can be used to assist the process of early diagnosis.

References

[1] Dora L, Agrawal S, Panda R, and Abraham A, 2018, *Nested cross-validation based adaptive sparse representation algorithm and its application to pathological brain classification*, Expert Syst. Appl., **114**, 313–321.
[2] Al-Tamimi M S H, and Sulong G, 2014, *Tumor Brain Detection Through Mr Images: A Review Of Literature*, J. Theor. Appl. Inf. Technol., **62(2)**.
[3] Özyurt F, Sert E, Avci E, and Dogantekin E, 2019, *Brain tumor detection based on Convolutional Neural Network with neutrosophic expert maximum fuzzy sure entropy*, Measurement, **147**, 106830.
[4] Tong J, Zhao Y, Zhang P, Chen L, and Jiang L, 2019, *MRI brain tumor segmentation based on texture features and kernel sparse coding*, Biomed. Signal Process. Control, **47**, 387–392.
[5] Shanthakumar P, and Ganeshkumar P, 2015, *Performance analysis of classifier for brain tumor detection and diagnosis*, Comput. Electr. Eng., **45**, 302–311.
[6] Sachdeva J, Kumar V, Gupta I, Khandelwal N, and Ahuja C K, 2016, *A package-SFERCB-"Segmentation, feature extraction, reduction and classification analysis by both SVM and ANN for brain tumors*, Appl. Soft Comput. J., **47**, 151–167.
[7] Abd-Ellah M K, Awad A I, Khalaf A A M, and Hamed H F A, 2019, *A review on brain tumor diagnosis from MRI images: Practical implications, key achievements, and lessons learned*, Magn. Reson. Imaging, **61**, 300–318.
[8] Mohan G, and Subashini M M, 2018, *MRI based medical image analysis: Survey on brain tumor grade classification*, Biomed. Signal Process. Control, **39**, 139–161.
[9] Kumar A, Ramachandran M, Gandomi A H, Patan R, Lukasik S, and Soundarapandian R K, 2019, *A deep neural network based classifier for brain tumor diagnosis*, Appl. Soft Comput. J., **82**, 105528.
[10] Öztürk Ş, and Akdemir B, 2018, *Application of Feature Extraction and Classification Methods for Histopathological Image using GLCM, LBP, LBGLCM, GLRLM and SFTA*, Procedia Comput. Sci., **132**, 40–46.
[11] Remeseiro B, and Bolon-Canedo V, 2019, *A review of feature selection methods in medical applications*, Comput. Biol. Med., **112**, 103375.
[12] Khalil M, Ayad H, and Adib A, 2018, *Performance evaluation of feature extraction techniques in MR-Brain image classification system*, Procedia Comput. Sci., **127**, 218–225.
[13] Arabi P M, Joshi G, and Vamsha Deepa N, 2016, *Performance evaluation of GLCM and pixel intensity matrix for skin texture analysis*, Perspect. Sci., **8**, 203–206.
[14] Abd Latif M H, Md Yusof H, Sidek S N, and Rusli N, 2015, *Implementation of GLCM Features in Thermal Imaging for Human Affective State Detection*, Procedia Comput. Sci., **76**, 308–315.
[15] Sharma K, Kaur A, and Gujral S, 2014, *Brain tumor detection based on machine learning*

algorithms, Int. J. Comput. Appl., **103 (1)**.
[16] Sudha B, Gopikannan P, Shenbagarajan A, and Balasubramanian C, 2014, *Classification of Brain Tumor Grades using Neural Network*, in Proceedings of the World Congress on Engineering 2014, **1**.
[17] Demirhan A, Törü M, and Güler I, 2014, *Segmentation of tumor and edema along with healthy tissues of brain using wavelets and neural networks*, IEEE J. Biomed. Heal. informatics, **19(4)**, 1451–1458.
[18] Gaikwad S B, and Joshi M S, 2015, *Brain tumor classification using principal component analysis and probabilistic neural network*, Int. J. Comput. Appl., **120(3)**.
[19] Vidyarthi A, and Mittal N, 2015, *Performance analysis of Gabor-Wavelet based features in classification of high grade malignant brain tumors*, in 2015 39th National Systems Conference (NSC), 1–6.
[20] Kumar S P M, and Chatteijee S, 2017, *Computer aided diagnostic for cancer detection using MRI images of brain (Brain tumor detection and classification system)*, 2016 IEEE Annu. India Conf. INDICON 2016.
[21] Banday S A, and Mir A H, 2016, *Statistical textural feature and deformable model based MR brain tumor segmentation*, 2016 Int. Conf. Adv. Comput. Commun. Informatics, ICACCI 2016, 657–663.
[22] R M A, and Scholar P G, 2017, *Brain Tumor Segmentation And Classification using DWT , Gabour Wavelet And GLCM*, 1744–1750.
[23] Islam A, Hossain M F, and Saha C, 2018, *A new hybrid approach for brain tumor classification using BWT-KSVM*, 4th Int. Conf. Adv. Electr. Eng. ICAEE 2017, 241–246.
[24] Sergaki E. *et al.*, 2018, *Application of ANN and ANFIS for detection of brain tumors in MRIs by using DWT and GLCM texture analysis*, IST 2018 - IEEE Int. Conf. Imaging Syst. Tech. Proc., 1–6.
[25] Dubey Y K, Mushrif M M, and Pisar *K, 2018, Brain tumor type detection using texture features in MR images*, IEEE Reg. 10 Humanit. Technol. Conf. R10-HTC, 1–4.
[26] Gopika G S, and Rajasree R S, 2018, *Certain Investigations on the Detection of Brain Tumors Using Support Vector Machine & Prinicipal Component Analysis*," in 2018 Fourth International Conference on Computing Communication Control and Automation (ICCUBEA), 1–5.
[27] Kim T K, 2019, *T test as a parametric statistic*, Korean J. Anesthesiol., **68(6)**, 540.
[28] Williams L L, and Quave K, 2019, *Comparing Two Groups: t-Tests*, Quant. Anthropol., 89–104.
[29] Feng Y, Huang Y, and Ma X, 2017, *The application of Student's t -test in internal quality control of clinical laboratory*, Front. Lab. Med., **1(3)**, 125–128.
[30] Smalheiser N R, 2017, *Null Hypothesis Statistical Testing and the t-Test*, Data Lit., 127–136.
[31] Buas M F, Li C I, Anderson G L, and Pepe M S, 2018, *Recommendation to use exact P-values in biomarker discovery research in place of approximate P-values*, Cancer Epidemiol., **56**, 83–89.
[32] Bergamelli M, Bianchi A, Khalaf L, and Urga G, 2019, *Combining p-values to test for multiple structural breaks in cointegrated regressions*, J. Econom., **211(2)**, 461–482.
[33] Jusman Y, Ng S C, Hasikin K, Kurnia R, Abu Osman N A, and Teoh K H, 2017, *A system for detection of cervical precancerous in field emission scanning electron microscope images using texture features*, J. Innov. Opt. Health Sci., **10 (2)**, 1650045.
[34] Jusman Y, Ng S C, Hasikin K, Kurnia R, Abu Osman N A, and Teoh K H, 2016, *Computer-aided screening system for cervical precancerous cells based on field emission scanning electron microscopy and energy dispersive x-ray images and spectra*, Opt. Eng., **55(10)**, 103110.
[35]Y. Jusman, S.-C. Ng, and K. Hasikin "Performances of Proposed Normalization Algorithm for Iris Recognition," Int. J. Adv. Intell. Informatics, Vol. 6, no. 2, July 2020.
[36]Y. Jusman, L. A. Dewiprabamukti, A. N. N. Chamim, Z. Mohamed, S. N. A. M. Kanafiah, & N. H. A. Halim. "Application of Watershed Algorithm and Gray Level Co-Occurrence Matrix in Leukemia Cells Images". IEEE. Paper presented at the 2020 3rd MECnIT. pp. 9-14, 2020
[37] Jusman Y, Tamarena R I, Puspita S, Saleh E, and Kanafiah S N A M, 2020, *Analysis of Features Extraction Performance to Differentiate of Dental Caries Types Using Gray Level Co-occurrence Matrix Algorithm*, Paper presented at the 2020 10th IEEE ICCSCE, 148-152.

IOP Conf. Series: Materials Science and Engineering 1082 (2021) 012012 doi:10.1088/1757-899X/1082/1/012012

Performance Measurement Analysis Using The Balanced Scorecard Method for Government Institution in Aceh

Raihan Dara Lufika[1], Syahriza[1], Cut Indrian Sary[1]
[1]Industrial Engineering Department, Universitas Syiah Kuala, Darussalam, Banda Aceh, Indonesia

E-mail: raihandlufika@unsyiah.ac.id

Abstract. The public and non public company have the same goals in adjusting their technology and system for competition. The performance measurement is one way to control the company to achieve their vision and mission. This research aims to determine whether or not the performance of a government institution in Aceh is according to their standard. The first step is identify the problems using the Balanced Scorecard method, with four perspectives, namely financial, customer, internal business processes, and growth and learning. The result shows that there are 18 Key Performance Indicators (KPI). The next step is to calculate the values of each indicators. Finally, the result is compared with the KPI standard. The final result shows that the lowest value of indicator in 2017 is the simple work procedures with a score of 47 and a final score of 0.015. Meanwhile, the highest value of indicator in 2017 is the budget utilisation with a score of 124 and the final score of 0.245. Overall, the Disperindagkop has a very low value of 13% which means the condition is very bad. Therefore, the company needs further improvements to increase their overall performance value.

1. Introduction

The globalisation era force company to adjust themselves with the latest technology and management system. Each company has their own values, visions, and missions. It is necessary for the company to measure the productivity rate of their employees to ensure its within the same scope. One of the most recommended and sufficient framework to manage firm is the balanced scorecard [1].

Balanced scorecard has been used as a tools for management and business for over 20 years [2]. The concept of balanced scorecard is flexible to use in the wide range of areas. The researches show that the balanced scorecard is beneficial to measuring the performance index in the health care service [3] and the academic sector such as polytechnic in Malawi [4] and higher education in Malaysia [5]. It shows that the balanced scorecard is not rigid to profit-oriented business only but useful for other form of organisation and company as well.

There are four perspectives in the balanced scorecard framework. Every part of the company works and contribute in harmony for reaching the company's goals with the adjustment of balance scorecard accordingly [6]. The case study for this research take places in one of the Government Institution in Aceh, we named it A company in this paper. Based on the observation, there are several problems in this institution such as employees arrived late, no regular meetings, untidy work management, messy parking area, not clear space partition, and messy documents management.

This study aim to measure the performance of this institution using the Balanced Score Card and provide possible improvement based on the analysis.

Content from this work may be used under the terms of the Creative Commons Attribution 3.0 licence. Any further distribution of this work must maintain attribution to the author(s) and the title of the work, journal citation and DOI.
Published under licence by IOP Publishing Ltd

2. Methodology

2.1. Data Collection

The data consists of two main data. The primary data is the questionnaire that distributed to 18 experts in the department. They are the high-level employee who knows the real situation in the department. The secondary data are other related documents such as the department hierarchy, the employee schedule, and others.

2.2. Data Processing

There are several process on this research including:

1. Determine of strategic plan based on four perspective of Balanced Scorecard
2. Determine of Key Performance Index (KPI) for each components on four perspective of Balanced Scorecard
3. Calculation of the Balance Scorecard

The result of the Balance Scorecard is the reflection of institution's performance. The analysis aims to understand which performance need an improvement. The last step will be a proposed strategic to improve the poor performance in the company.

3. Results and discussion

3.1. Strategy Map

Table 1 shows the strategic map of the department. There are four perspectives on the Balanced Scorecard which are financial, customers, internal business process, and learning and growth. For each perspective, there will be different number of components distinct to the situation in the department.

Table 1. Strategy Map of A company

Vision :
Creating a safe, prosperous, and independent society based on the UUPA through enhancement guidance for cooperatives and industry.
Missions :
1. Improve the service quality of the employees through the reformation of efficient, clean, orderly and authoritative bureaucratic 2. Accelerate economic growth and reduce poverty by maximising the utilization of regional resource potential and expanding employment opportunities 3. Increase the quality of education and health services 4. Increase the distribution of infrastructure to support basic services and the economy in an integrated manner 5. Create life that based on empowerment and social protection which values the Islamic law.
Goals and Objectives :
1. Develop a conducive business climate and partnerships 2. Optimizing the potential of local and international markets through increase competitiveness by utilise the export product market through quality control by institution 3. Competency-based training 4. Increase supervision, protection and law enforcement of occupational safety and health 5. Increase the number of new business units and various products in the industrial sector 6. Improve the quality of export goods through goods testing and quality certificates.

Financial	1	Budget growth
	2	Budget Utilisation
Customer	3	Customer Service
	4	Simple Registration Procedure
	5	Information Access
	6	Resolve Complaints
Business Process	7	Work Facilities
	8	Simple Work Procedure
	9	Activity According to Company's Targets
	10	Evaluation and Continues Improvement on Work Program
	11	Information Access
Learning and Growth	12	Employee Training
	13	Number of Employee
	14	Number of Resigned Employee
	15	Number of SMEs
	16	Technology Procurement and Utilisation
	17	Work Procedure on Latest Technology
	18	Compensation and Appreciation

3.2. Key Performance Index (KPI) values

The general manager with the approval of the head of department determine the Key Performance Index (KPI) values and targets. The values are the reference of the performance measurement. The highest the manager set the values standard, the more important that KPI for the company, vice versa.

Table 2. Key Performance Index (KPI) Values (percentage).

Perspectives	No.	Indicators	Source	Value	Value (%)
Financial	1	Budget growth	Company	0.184	18.4%
	2	Budget Utilisation	Company	0.198	19.8%
Financial Perspectives Value				0.382	38.2%
Customer	3	Customer Service	Survey	0.056	5.6%
	4	Simple Registration Procedure	Survey	0.043	4.3%
	5	Information Access	Survey	0.052	5.2%
	6	Resolve Complaints	Survey	0.055	5.5%
Customer Perspectives Value				0.206	20.6%
Business Process	7	Work Facilities	Company	0.045	4.5%
	8	Simple Work Procedure	Survey	0.032	3.2%
	9	Activity According to Company's Targets	Company	0.045	4.5%
	10	Evaluation and Continues Improvement on Work Program	Survey	0.042	4.2%
	11	Information Access	Survey	0.042	4.2%
Business Process Perspectives Value				0.206	20.6%

Learning and Growth	12	Employee Training	Company	0.032	3.2%
	13	Number of Employee	Company	0.035	3.5%
	14	Number of Resigned Employee	Company	0.025	2.5%
	15	Number of SMEs	Company	0.025	2.5%
	16	Technology Procurement and Utilisation	Survey	0.025	2.5%
	17	Work Procedure on Latest Technology	Survey	0.032	3.2%
	18	Compensation and Appreciation	Survey	0.032	3.2%
Learning and Growth Perspectives Value				0.206	20.6%
Total				1	100%

3.3. Balance Scorecard calculation result

For a financial perspective and growth and learning perspective, the data comparison is from the institution's performance from 2016 to 2017. For the customer perspective and the perspective of internal business processes, the questionnaire instrument is the tools for assessing the performance of A company's employees. Table 3 shows the performance calculation results.

Table 3. Performance measurement calculation result.

No	Perspectives	Indicators	KPI Values	Percentage	Target (%)	Realisation (%)	Score	Final Score	Percentage	Means
1	Financial	Budget growth	0.184	18.4	85	85	100	0.184	18.4	Very Good
		Budget Utilisation	0.198	19.8	80	99	124	0.245	24.5	Very Good
Total Values				38.2					42.9	
2	Customer	Customer Service	0.056	5.6	5.6	4.7	84	0.047	4.7	Good
		Simple Registration Procedure	0.043	4.3	4.3	3.0	79	0.030	3.0	Good
		Information Access	0.052	5.2	5.2	2.5	48	0.025	2.5	Bad
		Resolve Complaints	0.055	5.5	5.5	3.8	69	0.038	3.8	Enough
Total Values				20.6					14.0	
3	Business Process	Work Facilities	0.045	4.5	4.5	3.2	71	0.032	3.2	Good
		Simple Work Procedure	0.032	3.2	3.2	1.5	47	0.015	1.5	Bad
		Activity According to Company's Targets	0.045	4.5	4.5	4.1	91	0.041	4.1	Good
		Evaluation and Continues Improvement on Work	0.042	4.2	4.2	3.7	88	0.037	3.7	Good

		Program								
		Information Access	0.042	4.2	4.2	3.9	93	0.039	3.9	Good
Total Values				20.6					16.4	
4	Learning and Growth	Employee Training	0.032	3.2	65	63	97	0.031	3.1	Very Good
		Number of Employee	0.035	3.5	66	66	100	0.035	3.5	Very Good
		Number of Resigned Employee	0.025	2.5	3 persons	3 persons	100	0.025	2.5	Very Good
		Number of SMEs	0.025	2.5	50 units	25 units	50	0.013	1.3	Enough
		Technology Procurement and Utilisation	0.025	2.5	90	75	83	0.021	2.1	Good
		Work Procedure on Latest Technology	0.032	3.2	90	47	52	0.017	1.7	Enough
		Compensation and Appreciation	0.032	3.2	60	60	100	0.032	3.2	Very Good
Total Values				20.6					17.3	
Total			1	100				0.906	90.68	

Two indicators labelled bad which are information access and simple work schedule. For the enough label, there are three indicators including in this label. Then, there are seven indicators in label good. Finally, the indicators for label very good are as much as six indicators.

3.4. Conversion of Final Performance Values Calculation

The final calculation values of performance measurement is the initial value for calculating the overall performance value. The calculation of overall performance value refers to Likert Scale. Table 4 shows the conversion result.

Table 4. Conversion of final performance value calculation.

Perspective	Indicators	Criteria	Score
Financial	Budget growth	Very Good	5
	Budget Utilisation	Very Good	5
Customer	Customer Service	Good	4
	Simple Registration Procedure	Good	4
	Information Access	Bad	2
	Resolve Complaints	Enough	3
Business Process	Work Facilities	Good	4
	Simple Work Procedure	Bad	2
	Activity According to Company's Targets	Good	4

	Evaluation and Continues Improvement on Work Program	Good	4
	Information Access	Good	4
Learning and Growth	Employee Training	Very Good	5
	Number of Employee	Very Good	5
	Number of Resigned Employee	Very Good	5
	Number of SMEs	Enough	3
	Technology Procurement and Utilisation	Good	4
	Work Procedure on Latest Technology	Enough	3
	Compensation and Appreciation	Very Good	5
Total Score			71
Total Respondent			18
Higher Score			540
Lowest Score			72
Total Score Percentage			13%
Means			Very Bad

The determination of the calculated performance score's weight is using several elements, namely, KPI weight, target value, and year-end realization. The scoring item refers to the Likert scale. The highest Likert scale value is 124 divided by five scales. The scale value is 24.8. The Likert scale value from 1 to 5. Value 1 is very bad (0-24.99%), value 2 is bad (25-49.99%), value 3 is enough (50-74, 99), score 4 is good (75-99.99%), and score 5 is very good (100-124%).

3.5. Proposed Improvement

Simple work procedures is an indicator that has the lowest score and realization value. The correction planning is to have concrete remedial steps and clear work description in the form of chart, map or simulation image. Information access also has a low score. It is only one point different from the score of simple work procedures which are 48 and 47 respectively. The future suggestions are to do the initial repairs for website to access update information.

Resolve complaints gets a score of 69 out of a total score of 124. Therefore, the service is labelled enough. The improvement of this indicator is not necessary but need consideration. The suggestions for improvements are to increase special services for resolving community complaints, accommodating all complaints in the quickest best way, service for every public complaint should be transparent. Indicator with the name of number of SMEs gets a score of 50 out of a total score of 124, so the service is enough. It is necessary to increase the number of SMEs by providing services and regular skills training for them. The work procedure on latest technology has labelled enough as well. The future suggestion for this indicators is to upgrade the technological equipment and software in the institution.

4. Summary

In conclusion, the institution has a 'very bad' performance overall. The indicators for this study is eighteen in total. After the calculation, there are two, three, seven, and six indicators labelled bad, good, enough, and very good respectively. Considering the result, we only suggest the corrective plan for indicators with label bad and good. The suggestion for simple work procedures is to have concrete remedial steps and clear work description in the form of chart, map or simulation image. Then for information access, we suggest doing the initial repairs for the website to access update information. Next, the suggestions for resolve complaints are to increase special services for resolving community complaints, accommodating all complaints in the quickest best way, service for every public complaint should be transparent. After that, to improve the number of SMEs, it is necessary to increase the services and regular skills training for them. Finally, the future suggestion for work procedure on the latest technology is to upgrade the technological equipment and software in the institution.

References

[1] Kopecka N, 2015, *The Balanced Scorecard Implementation, Integrated Approach, and the Quality of Its Measurement*, Procedia Economics and Finance, **25**, 59-69.
[2] Madsen D O, and Stenheim T, 2015, *The Balanced Scorecard: A Review of Five Research Areas*, American Journal of Management, **15(2)**, 24-41.
[3] Suprapto B, Wahab H A, and Wibowo A J, 2009, *The Implementation of Balance Score Card for Performance Measurement in Small and Medium Enterprises: Evidence from Malaysian Health Care Services*, The Asian Journal of Technology Management, **76-87**
[4] Chimtengo S, Mkandawire K, and Hanif R, 2017, *An evaluation of performance using the balanced scorecard model for the university of Malawi's polytechnic*, African Journal of Business Management, 84-93.
[5] Binden W, Mziu H, and Suhaimi M A, 2014, *Employing the Balanced Scorecard (BSC) to Measure Performance in Higher Education – Malaysia*, International Journal of Information and Communication Technology Research, **4(1)**, 38-44.
[6] Chavan M, 2007, *The balanced scorecard: a new challenge*, Journal of Management Development, **28(5)**, 393-406.

Mechanical and Acoustical Properties of Short Oil Palm Frond Fibre Reinforced Polyester Composites

Sabri[1,2]*, Zahrul Fuadi[1,2], Rudi Kurniawan[1,2], Samsul Rizal[1,2]

[1]Doctoral Program, School of Engineering, Universitas Syiah Kuala, Banda Aceh, Indonesia, 23111
[2]Department of Mechanical Engineering, Universitas Syiah Kuala, Jln. Syech Abdurrauf No.7 Darussalam – Banda Aceh, Indonesia, 23111

*E-mail: sabri@unsyiah.ac.id

Abstract. This paper presents the determination of tensile, flexural, and acoustical properties of short oil palm frond (OPF) fibre reinforced polyester composites and issues that have to be controlled in order to improve those properties. Three composite samples were fabricated using fibre content of 5%, 10% and 15% (by mass fraction) each. The tensile strength and elastic modulus of the composites were determined by tensile test conforming to the ASTM D638 standard and the flexural strength was measured by the three-point bending test method in accordance to ASTM D790 standard whereas the acoustical properties were determined by transfer function method conforming to the ISO 10534-2 standard. It is showed that the OPF-polyester composites possess greater tensile strength and flexural strength at lower fibre content while the elastic modulus registered increment with higher fibre loadings. Reduction in tensile and flexural strengths at higher fibre content is caused by poor stress transfer mechanism as the combined effect of inadequate fibre/ matrix interface adhesion and stress concentration spots. Fibre agglomeration, which causes hindrance to the polyester macromolecule chain mobility, is accounted for the increment in stiffness properties as the amount of fibre is increased. From the measurement of the acoustical properties shows that the sound absorption coefficient is a function of incidence frequency. The normal incidence sound absorption coefficients for different frequencies tend to increase with increasing frequency operation. There is an excellent agreement between the absorption coefficient and the input frequency.

Content from this work may be used under the terms of the Creative Commons Attribution 3.0 licence. Any further distribution of this work must maintain attribution to the author(s) and the title of the work, journal citation and DOI.
Published under licence by IOP Publishing Ltd

1. Introduction

Composites are a new class of material made from the combination of different basic resources such as metals, plastics, glass or ceramics. The objective of this coalition is to gain the synergy from the components to form a new material that has superior performance than the individual components in terms of mechanical and acoustical properties, performance to weight ratio and feasibility of process. Among the possible combinations, the utilization of plastics as the matrix reinforced with synthetic fibres such as glass mat or carbon is more conventional. Commodity thermoplastics such as polyethylene (PE), polystyrene (PS), poly-vinyl-chloride (PVC) and polypropylene (PP) are commonly reinforced with glass-fibre-mat (GMT) for various applications and have been dominion in the plastics market for years [1]. Thermoset plastics such as polyester and epoxy also have wide application domains due to its high viscosity characteristics, suitable to be reinforced with intermediate and long fibres using hand lay-up fabrication method [2]. Elastomers like natural rubber (NR) blended with thermoplastics have been extensively utilized for research purposes [3,4].

Natural fibre, as in composite application for mechanical and structural engineering, can be defined as reinforcing agents extracted from naturally occurring resources such as agricultural dregs or agriculture based industrial residues. Natural fibres therefore also called agro-based fibres. These materials are in their own composites made from lignin matrix bounded by cellulose fibres [5].

There is a wide variety of agro-based fibres and they should be considered for composites fabrication to take advantage of the unique fibre properties, local source of availability and cost. Flax, bamboo, rice husk, wood flour, oil palm frond and empty fruit bunch (EFB), sugarcane and coir are among the popular choices found in the literature [1,4,6]. The serious consideration given to these potential alternatives are due to the inherent drawbacks in the processing, utilization and disposal of the synthetic fibres' composites.

The main advantages of using lignocelluloses fibres are its low density, superior specific properties, environmental friendly, greater fibre loading possibility, variety of fibres available, low energy consumption in extraction, low decomposition rate and show less concern with personal safety and health. Besides that, the fibres are non-abrasive hence minimizes processing tool impairment in short term. In fact, the downstream processing of waste natural resources into value-added products will generate extra income for the plantation sector.

2. Theoretical Consideration

Palm trunks and fronds are obtained from the plantation ground whereas EFB, shells and mesocarp fibres are common residues found in the mills [3]. Table 1 below shows the nominal fibre dimensions of oil palm trunk, frond and bunches.

Table 1. Fibre dimension of the oil palm trunk, frond and bunch [7].

Dimension	Trunk	Fronds	Bunches
Length, mm	1.22	1.52	0.89
Width, µm	35.3	-	25.0
Thick, µm	4.5	-	2.8

Oil palm by-products are usually converted into value-added products in the downstream processes. The attempt to renovate these by-products into useful raw materials had been initiated since the 1970's. Peh *et al.* (1976) had studied the possibility of producing pulps from EFB by chemical-modification means. Khoo and Lee (1991) have described the methods of extracting the oil palm trunks, fronds and EFB to manufacture pulp and paper. Besides this, oil palm by-products have also been utilized to produce particleboards, cement tiles, reconstituted boards, chemical and fuel. The availability, potentials and consistency in supply of the oil palm by-products are crucial to sustain the relevance of oil palm lignocelluloses residues to those industries.

Nicolais and Nicodemo (1973) suggested the following equations which relate the tensile strengths of the composite to the matrix:

$$\sigma_c = \sigma_p \,(1 - a\phi^b) \tag{1}$$

where ϕ is the fibre volume fraction and constants a and b represent the stress concentration factor and filler geometry, respectively.

Elastic modulus of a two-phase composite can be predicted using the Einstein's equations:

$$E_r = 1 + \phi \tag{2}$$

$$E_r = 1 + 2.5\phi \tag{3}$$

where E_r is the composite's modulus (E_c) relative to the matrix's modulus (E_p). Equation (2) is deemed for the case of non-adhesion at the fibre/matrix interface while Equation (3) is considered if a good adhesion at the interface is expected by means of fibre surface modification or at the presence of a coupling medium. Haji A. *et al.* (2005) cited another formulation which accounts for planar fibres as given below:

$$E_c = E_p\,(1 - \phi) + E_f.\phi\ .\ MRF \tag{4}$$

MRF in equation (4) represents the modulus of reduction factor that depends on the aspect ratio of the flakes and percentage of fibre loadings.

The measurement of normal incidence absorption coefficient of materials using the two-microphone method or transfer function method is shown schematically in figure 1. A sample of the material to be tested is placed in a sample holder and mounted to one end of a straight tube. A rigid plunger with an adjustable depth is placed behind the sample to provide a reflecting surface. A sound source, typically a high-output acoustic driver, is connected at the opposite end of the tube. A pair of microphones is mounted flush with the inner wall of the tube near the sample end of the tube.

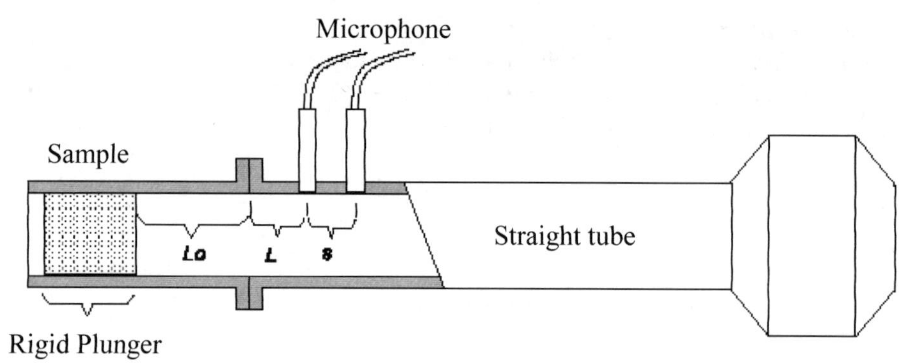

Figure 1. The two-microphone impedance measurement tube testing according to ISO 10534-2.

A multi-channel spectrum analyzer is used to obtain the transfer function (frequency response function) between the microphones. In this measurement, the microphone closer to the source is the reference channel. From the transfer function H_{12}, the pressure reflection coefficient R of the material is determined from the following equation:

$$R = \frac{H_{12} - e^{-jks}}{e^{jks} - H_{12}} e^{j2k(L+s)} \tag{5}$$

where L is the distance from the sample face to the first microphone and s is the distance between the microphones, $k = 2\pi f/c$, f is the frequency, and c is the speed of sound. From the reflection coefficient, the absorption coefficient α of the sample may be determined from the following equation:

$$\alpha = 1 - |R|^2 \tag{6}$$

3. Materials and Method

Raw oil palm (*Elaeis guineensis*) fronds were initially threshed into smaller sections using the RLL-3 Portable Crusher. Fronds, which have harder skin, were slightly hammered and teased in order to gain the fibre clumps and strands insight. The fibrous strands obtained were washed to remove the dirt and unwanted matters. The strands were then dried in the oven at 363 K for about 24 hours before crushing them to obtain short oil palm frond fibres (OPF). The nominal diameter of the fibre is 30 μm and 0.8 mm in length.

Revelsol P9509 unsaturated polyester was chosen as the matrix. Three samples were fabricated based on different fibre mass fraction (m_f) i.e. 5 %, 10 % and 15 %. Another sample consisting 100 % polyester was also manufactured for comparison purposes. The samples were fabricated using the hand lay-up method collectively with hot pressing technique so as to obtain specimens with even surface and thickness. Appropriate amount of polyester and OPF, accordingly to the mass fraction, were first blended manually and 2 ml of hardener (Butanox M50 Methyl-Ethyl-Cetone-Peroxide, 33 % in dimethyl phthalate) were mixed utterly into the former mixture. The final mixture was left in the ambient temperature for 1 minute to coagulate and then released into the dumbbell shaped mould. The mould was pressed at 30 MPa, 301 K for nearly 4 hours using GOTECH Hot Press Machine. The fabrication process was reiterated with the fibre fractions stated above.

Tensile tests were performed on the samples in conformance to the ASTM D638 standard. The Universal Testing Machine 10 kN, Testometric Micro 500 (UK) were used for the tests. Three specimens from each sample were tested at a strain rate of 1 mm/min. The flexural strength of the samples was determined by the flexural test, which was done according to the ASTM D790 – 92 standard, Test Method in. Figure 2 illustrates the experimental set-up. The support-span to depth ratio was set as 3:1. The nominal length and thickness of the specimens were 150 mm and 3 mm, respectively. Rate of cross-head motion was 5.3 mm/min.

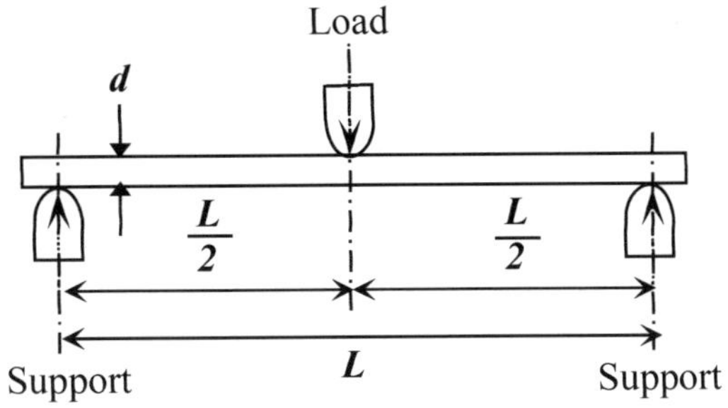

Figure 2. Three-point loading system.

The process of removing sound power from a particular space in lowering the sound pressure level within a particular volume is known as *Sound Absorption*. This process is concerned with a lowering of the energy content of the pulsating air as it encounters a porous or flexible medium, it should be recognized as being completely different from most other type of "absorption". A blotter, for example, can absorb liquids when only a corner is immersed into the fluid, but sound absorbing materials cannot pull sound out of the air.

Sound is a source of energy which can only be reduced when it is actively directed into and through a medium in which it experiences frictional resistance, in which its kinetic energy is converted to heat. There is an optimal airflow resistance for the absorption of sound at any particular frequency. A very porous material with a very low airflow resistance is the common household furnace filter. Sound will pass through such a filter with negligible reduction. At the other extreme, if a material has a very high flow resistance, the sound cannot readily penetrate to be absorbed and must of necessity reflect back and add energy to the space from which it came. Sound absorption coefficient is a fraction of the incident sound energy absorbed by a material of any size. It can be measured using impedance tube as shown in figure 3, based on the transfer function between two microphones.

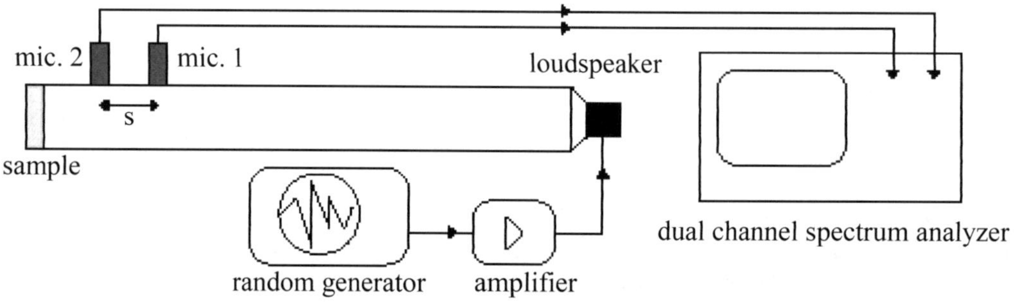

Figure 3. Schematic diagram for measuring sound absorption coefficient in impedance tube

4. Results and Discussion

Table 2 shows the results of the tensile and three-point flexural tests. Elastic modulus of the samples increased as the fibre mass fraction is greatened. An optimum increment of 77 % was registered for the 10 % m_f sample as compared to the plain (100 %) polyester. Enhancement of the stiffness properties with fibre addition is, fundamentally, due to the effect of fibre agglomeration in the polyester matrix. Fibre addition into the polyester causes obstruction to the polyester macromolecule chain mobility.

Table 2. The tensile and flexural properties of short oil palm frond fibre reinforced polyester.

Fibre Mass Fraction (%)	Tensile Test		Flexural Test
	Tensile Strength (MPa)	Elastic Modulus (GPa)	Flexural Strength (MPa)
0	16.50	2.11	65.7
5	15.20	3.61	25.6
10	8.00	3.73	25.3
15	7.10	3.70	20.1

The partiality in the dispersion of the fibre is the main factor that restricts the mobility of the chain. Short fibres, which are naturally fine and have small surface area, tend to form fibre aggregates by way of hydrogen bonding at the hydrophilic surface of the fibre. Subsequently, a group of aggregates, of its own accord, fuse to form a larger structure called agglomerate [6]. It is this agglomeration effect that hinders the polymer chain mobility. This also causes the elongation at break to be reduced. Since the elongation is limited, the ductility of the samples is affected resulting in a more rigid material.

The tensile strength of the composites, as shown in Table 2, was found to be somewhat inversely proportional to the fibre loading. Among the factors that influence the reduction in tensile strength is the emergence of stress concentration points arising from weak fibre/matrix interface adhesion. Besides that, deterioration of the crystallinity of the matrix in the presence of fibre could also be the reason for the decrement in the tensile strength. The fibre dispersion, which is not equal in all directions, develops agglomeration spots where phase discontinuities occur (figure 4).

Discontinuities can also crop up between fibres of different geometry. These discontinuities impede the matrix-to-fibre stress transfer mechanism hence causing applied stresses to concentrate at the fibre/matrix interface. This in turn prompted the composites to fail at an early stage of extension. One prevalent remedy to reduce the agglomeration effect and stress concentration spots is to chemically modify the fibre surface using suitable coupling agent, silane for instance. The coupling agent acts to modify the fibre surface by forming *organosilanol* and *siloxanes* groups to promote the interface adhesion, which eradicates, to the minimum level, stress concentration points in the composites [2].

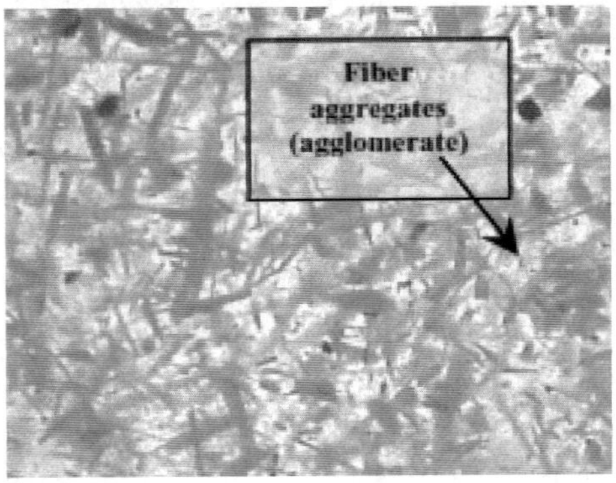

Figure 4. Photograph of the fibre agglomeration in the composite.

Increment in fibre mass fraction also leads to a reduction in flexural strength as shown in figure 5. Compared to the reduction in the tensile strength of only 8 % from 0 % to 7 % m_f samples, the flexural strength reduced 60 % for the same range. However, the values of the flexural strength remained superior to those of tensile strength. Thus, it is apparent that OPF/polyester composites are stronger in bending mode than in tensile mode.

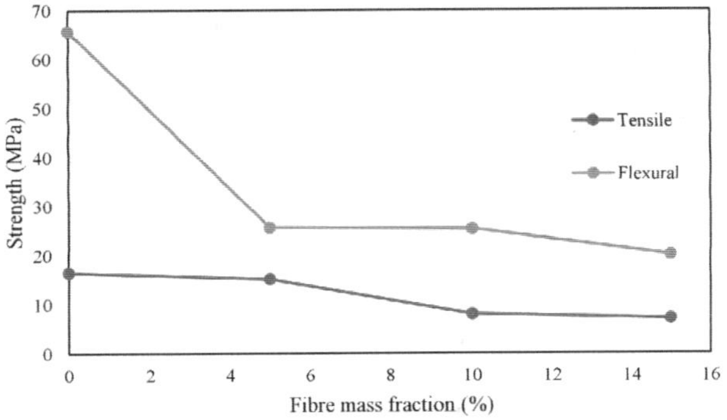

Figure 5. Tensile and flexural strength of OPF/ polyester composite.

From the measurement of the acoustical properties show that the sound absorption coefficient is a function of incidence frequency. The normal incidence sound absorption coefficients for different frequencies tend to increase with increasing operation frequency. Figure 6 shows that the normal incidence sound absorption coefficients for working frequency range from 125 – 4000 Hz.

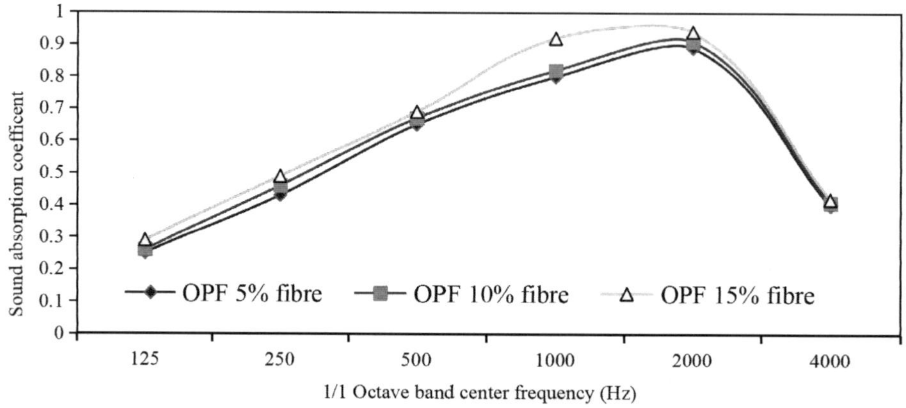

Figure 6. Sound absorption coefficient for OPF/ polyester composite.

As can be seen, there is an excellent agreement between the absorption coefficient and the input frequency, especially for frequency below 2000 Hz, but unfortunately, the test materials exhibit low absorption at frequency upper than 2000 Hz.

5. Conclusions

The tensile and flexural properties of OPF/polyester composites were determined. It has been observed that the modulus of elasticity is directly proportional to the fibre mass fraction in the composite whereas strength, in both the tensile and bending mode, are reduced by fibre addition. Hence, higher amount of fibre loadings produces stiffer but weaker composites. The hydrophilic trait of the fibres in tandem with its high moisture content creates agglomeration issues whereby the fibre/matrix interface bond is affected. Although this yields a rigid composite, the formation of stress concentration points in fact diminishes the composite's strength. Thus, the overall mechanical performance of the composite system is ought to be controlled by the strong interface bonding and uniform dispersion of the fibres. In order to gain these qualities, fibre surface modification is indispensable. Utilization of coupling agent, such as silane, that facilitates the development of a sturdy bond between the hydrophilic fibre and hydrophobic matrix is recommended in the literature. From the measurement of the acoustical properties show that the sound absorption coefficient is a function of incidence frequency. The normal incidence sound absorption coefficients for different frequencies tend to increase with increasing frequency operation. There is an excellent agreement between the absorption coefficient and the input frequency.

Acknowledgments

This publication was made possible by a grant from the Research Fund of the Ministry of Research, Technology, and Education, Republic of Indonesia, and LPPM Universitas Syiah Kuala; the financial support is greatly appreciated. We would like to thank Mr. Bakri for their support during this experiment.

Reference

[1] Chauhan V, Kärki T, and Varis J, 2019, *Review of Natural Fiber Reinforced Engineering Plastic Composites, Their Applications in The Transportation Sector and Processing Techniques,* Journal of Thermoplastic Composite Materials, **41**, 0001 – 0041, DOI: http://dx.doi.org/10.1177/0892705719889095.
[2] Kicki F K, and Astrom B T, 1997, *Manufacturing and Applications of Structural Sandwich Components*. Composites Part A, **28(2)**, 97 – 111, https://doi.org/10.1016/S1359-835X(96)00098-X
[3] Kamaruzzaman S, Rozli Z, Sahari, and Othman, 1998., *Thermal Performance of Thermoplastic Natural Rubber Solar Collector*. AMPT '98, **6**, 247 – 253.
[4] Kobayashi Y et al, 2015, *Thermomechanical Pulping and Its Application to Empty Fruit Bunches of Oil Palm.* Proc. National Symposium on Oil Palm By-prod. For Agro-based Industries, **11**, 67 – 78.
[5] Heru S, Eko M, Yudy S I, and Rudy S, 2014, *Morphology, Structure, and Mechanical Properties of Natural Cellulose Fiber from Mendong Grass (fimbristylis globulosa)*, Journal of Natural Fibers, **11(4)**, 333 – 351, DOI: 10.1080/15440478.2013.879087.
[6] Maiti S N, and Singh K, 1986, *Influence of Wood Flour on the Mechanical Properties of Polyethylene*, Journal of Applied Polymer Science, **32**, 4285 – 4289.
[7] Roger M R, Anand R, Sanadi, Daniel F C, and Rodney E J, 1997, *Utilization of Natural Fibres in Plastic Composites*, http://www.fpl.fs.fed.us/documents/PDF1997/rowel97d.pdf.
[8] Peh T B, Khoo K C, and Lee T W, 1976, *Pulping Studies on Empty Fruit Bunches of Oil Palm,* The Malaysian Forester, **39(1)**, 22 – 30.
[9] Khoo K C, and Lee T W, 1991, *Pulp and Paper from the Oil Palm*, Appita, **44(6)**, 385 – 388.
[10] Ahmad F, Omar M Y, Ishak S, and Omar M A, 1994, *Application of Rice Husk Ash as Fillers in Polypropylene: Effect of Silane Coupling Agents*, Journal of Science Technical Dev, 115 – 140.
[11] Haji A, Abdullah, and Othman H, 2005, *Development of Short Fibre Thermoplastic Natural Rubber Composites*, Proc. ICAST'05, 251 – 255.
[12] Muhammad H, Zakaria, and Abdul H H, 2015, *Potentials of Oil Palm By-Products as Raw Materials for, Agro-Based Industries*, Proc. National Symposium on Oil Palm By-prod. For Agro-based Industries, 7 – 15.
[13] Jain S, and Kumar R, 1992, *Mechanical Behaviour of Bamboo and Bamboo Composite*, Journal of Materials Science, **27**, 4598 – 4604.
[14] Rozli Z, Fatt L K, Che H A, and Jailani S, 1998, *Effect of Natural Rubber on the Interlaminar Fracture Properties of Fibre Reinforced Thermoplastic Composites*, AMPT '98, 232 – 238.
[15] ASTM D638-14, 2014. *Standard Test Method for Tensile Properties of Plastics.*
[16] ASTM D790-99, 1999, *Standard Test Methods for Flexural Properties of Unreinforced and Reinforced Plastics and Electrical Insulating Materials.*
[17] ISO 10534-2, 2001, Acoustics - *Determination of Sound Absorption Coefficient and Impedance in Impedance Tubes* - Part 2: Transfer-function method.

AUTHOR INDEX